the
ILLUSTRATED ENCYCLOPEDIA OF WOODWORKING & HANDTOOLS INSTRUMENTS & DEVICES

also by Graham Blackburn

❦

Illustrated Housebuilding

Illustrated Basic Carpentry

The Postage Stamp Gazetteer

Illustrated Furniture Making

Illustrated Interior Carpentry

*The Illustrated Encyclopedia of Ships, Boats,
Vessels, & other Water-borne Craft*

The Illustrated Dictionary of Nautical Terms

The Parts of a House

An Illustrated Calendar of Home Repair

Quick & Easy Home Repair

Floors, Walls, & Ceilings

Creative Ideas for Household Storage

A Manual of Classic Woodworking Handtools

the

ILLUSTRATED ENCYCLOPEDIA OF
WOODWORKING &HANDTOOLS
INSTRUMENTS DEVICES

Revised Edition

CONTAINING A FULL DESCRIPTION OF

THE TOOLS

used by CARPENTERS, JOINERS, and CABINETMAKERS,

with many examples of tools used by other woodworkers such as
WOODSMEN, SAWYERS, COACHMAKERS, WHEELWRIGHTS, SHIPWRIGHTS,
WAINWRIGHTS, COOPERS, TURNERS, PATTERNMAKERS, and WHITTLERS.

WITH
MORE THAN *500* LINE DRAWINGS

written and illustrated by
Graham Blackburn

The Globe Pequot Press

Chester, Connecticut

Designed and set by Graham Blackburn

LIBRARY OF CONGRESS CATALOGING-IN-PUBLICATION DATA

Blackburn, Graham, 1940–
    The illustrated encyclopedia of woodworking handtools instruments
and devices: containing a full description of the tools used by carpenters,
joiners, and cabinetmakers . . . / written and illustrated by Graham
Blackburn.—Rev. ed.
        p   cm.
    ISBN 0-87106-168-6
    1. Woodworking tools—Dictionaries. 2. Woodworking tools—History.
    I. Title.
TT186.B52 1992
684'.082'028—dc20   91-25510
                                                              CIP

Manufactured in the United States of America
First Globe Pequot Edition/First Printing

for my parents
Verena Blackburn
John Blackburn

The Nail Puller, from *Elementary Carpentry and Joinery,* by F. Y., ca. 1870

ACKNOWLEDGMENTS

I would like to thank Cathy Maher for introducing me
to Julie D'Alton Houston, Julie for her help, Heide
Duggal for her encouragement, and Maria
Muldaur for her house.

PREFACE TO THE REVISED EDITION

In the seventeen years since this book was first published there has been a tremendous growth in the interest shown in old tools, handwork, and especially in custom-made furniture. This last—just one facet of the newly burgeoning crafts revival—has in turn focused many people's attention on the tools and techniques used by all branches of woodworking, from carpenters to cabinet-makers. Consequently, what was once a somewhat obscure field has become part of the mainstream for collectors and practitioners alike. There are now numerous journals and societies devoted to tool collecting. This, together with the fact that *The Illustrated Encyclopedia of Woodworking Handtools Instruments & Devices* received such favorable attention, has made possible a new edition, and I am pleased to be able to include many additions and emendations that have been brought to my notice over the years.

Apart from the new typeset look, the format has changed little: some of the more obvious cross-references have been omitted, and in a few cases the order has been changed for reasons of logic and clarity. All tools in *italics* may be looked up as a separate entry, and major groups are listed at the head of each section in the order they are covered, providing a convenient overview and making easier their final discovery.

The extra material has, I hope, made this a more useful work, but I shall continue to be grateful for errors and omissions brought to my notice.

Graham Blackburn
Point Reyes Station, California, 1991

INTRODUCTION

Scope

This book started out as an illustrated book of carpenters' tools. It then grew to include a few tools that, while not actually "carpentry" tools, might be found in the average carpenter's toolbox. By the time I had decided to include woodworking tools other than those that a carpenter might use—for example, tools used by cabinetmakers or joiners—it had become apparent that I would have to establish some limits, for the list was growing endlessly.

The first qualification for a tool's inclusion, therefore, was that it must be a "handtool." No matter how the other qualifications may have been occasionally disregarded, the requirement of a tool being a handtool was strictly observed. There are included in this book no power tools of any description, whether powered by water, gasoline, or electricity. Even foot-tools, such as treadle-operated jigsaws or lathes have been excluded.

The second major qualification was that the tool be used by woodworkers. This includes all those varied trades in which wood is the main material, such as those of not only the carpenter, but also, among others, the joiner, the cabinetmaker, the patternmaker, the cooper, the coachbuilder, and even the woodsman and the whittler. This qualification, however, has not been adhered to completely. When dealing with a class of tool such as the hammer, which, although much used in a variety of forms by woodworkers is also used by other trades, I have included examples of the non-woodworking types for the sake of completeness. I have also included a few tools with dubious woodworking connections simply because they appealed to me.

Although I have attempted to give some historical background, particularly with regard to the larger and older tool groups, such as adzes, axes, chisels, and planes, this is not a historical treatise on woodworking tools, but rather an index, as comprehensive as possible, of those tools common to western civilization from the eighteenth century to the present day. Few tools from the past

are missing, most having remained essentially the same for centuries. Furthermore, many that may have become obsolete and disappeared have usually left familiar descendants. The Roman carpenter might not recognize some of the newer tools in this book, but it is doubtful that he would miss many from his own toolbox.

Format

Although related tools, such as the various chisels and the many planes, are grouped together, the entire contents are listed alphabetically. If a particular tool is sought under its own name it will be found—with a cross-reference when necessary.

Where there are many examples of one class of tool, a comprehensive list will be found at the beginning of the group, as well as cross-references to each member within the work as a whole.

Any tool in *italics* (e.g., *maul*) may be looked up as a separate entry.

Additionally, many technical terms and operations have been explained and illustrated wherever it has been necessary to include them in the description of particular tools.

Aim

My aim has been to gather together in one place as many of the tools connected with woodworking as possible that are solely dependent on the strength and ingenuity of a single human being.

Wood is no longer the only important material in our society. It now takes its place along with plastics and metals, which require different tools and approaches to be worked. It is only natural that man should seek to apply efficient techniques developed through the working of one material to the working of another. In very

many cases the introduction of power tools has greatly benefited those trades and industries previously dependent on handtools. Though it is hard to imagine a spaceship being built solely with the tools of the carriagemaker, it would be equally hard to imagine a world bereft of the simpler beauty of a handcrafted piece of wooden furniture.

It is for this reason that I think the crafts of the premechanized age should be augmented, and not merely supplanted, by modern technology; it would be a pity to lose that body of knowledge represented by handtools in the name of economy or efficiency.

For those who may be interested in pursuing further either the historical aspect or the technical application of the tools listed in this book, a selective bibliography concerned primarily with woodworking handtools, as well as a list of some of the places where many of the older tools may be seen, are included at the end of the book.

Grave, or even slight omissions brought to my attention will be gratefully received, for if through the improvement of this book the survival of beauty and the identity of the individual tool can be furthered, we shall all benefit.

Hitherto I cannot learn that any hath undertaken this Task, though I could have wished it had been performed by an abler hand than mine; yet, since it is not, I have ventured upon it.

Joseph Moxon, 1703

From the preface to Joseph Moxon's book, *Mechanic Exercises or the Doctrine of Handy-Works*

The Carpenter, from *Das Ständebuch* (The Book of Trades), by Jost Amman, 1568

Being used, from top to bottom: two-man crosscut-saw, auger, twibill, axe, holdfast. Also visible in foreground: auger, axe, broadaxe, twibill, chalk line.

A

Adjustable Plumb, Level, and Inclinometer: see *Level*.

Adjustable Rabbet Plane: see *Plane, Special-purpose*.

Adze

This is a shaping and smoothing tool characterized by having its blade set at right angles to its haft (also called the helve or handle). It has been in use for thousands of years; the Egyptians thought so highly of this tool that in their picture-writing an adze meant "the able one." Since in use it is swung towards the user's legs, it has also earned the nickname "old shin splitter."

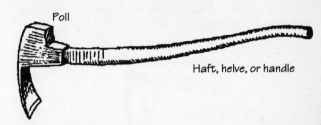

Poll

Haft, helve, or handle

• Adze, Carpenter's
Older carpenter's adzes have straight hafts. Note the square poll in the back for use as a *maul*.

• Adze, Cooper's (Howel)
A cooper's adze or howel is much smaller than a *carpenter's adze*. It is held in one hand, and has a very curved blade called a "colt's foot." Largely replaced by the *chiv plane*, it was used to prepare the inside ends of the barrel staves to receive the heads of the cask or barrel. The depression it makes is known as the howel.

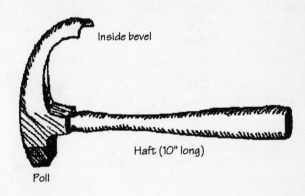

Inside bevel

Haft (10" long)

Poll

• Adze, Gutter
Also called a "spout adze." Early adzes were flat-headed (made without polls), and the hafts were generally

straight—without the curve that later became characteristic of most full-size adzes.

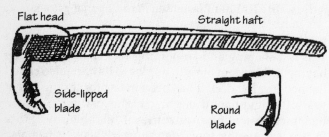

Flat head · Straight haft · Side-lipped blade · Round blade

• Adze, Sculptor's

Sculptor's adzes, like their close cousins, *cooper's adzes*, are small adzes made for use with one hand.

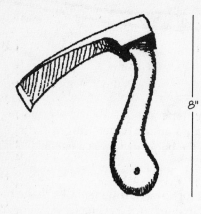

8"

• Adze, Ship Carpenter's

Also known as a "shipwright's adze," the ship carpenter's adze is much broader and flatter than other adzes, and is further characterized by the addition of a spur, used as a *nail set* or *punch*, on the back of the head. Like the *gutter adze* it may be round or side-lipped.

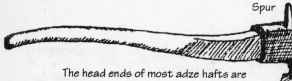

Spur

The head ends of most adze hafts are square, so that the head, which is fitted from the handle end, cannot fall off in use.

Angle Divider

With this tool any angle can be found and bisected. It is especially useful when fitting oddly-angled moulding or trimwork. The cross-piece at the bottom enables it to be used as a *square*.

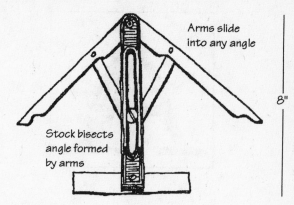

Arms slide into any angle · Stock bisects angle formed by arms · 8"

Astragal Plane: see *Plane, Moulding*.

Auger

This word comes from the Teutonic *nafugar*, which means literally "nave borer"—nave being the hole in the hub of a wheel through which the axle passes. Like many other words in English, such as *orange* and *adder*, the initial *n* has been lost. An auger is properly a complete tool in itself, with a transversely fitted handle, used for boring holes in wood. The word *auger*, however, is sometimes erroneously used to mean an *auger bit*, which is something used in a *bit brace*.

• Auger, Burn

This tool, being heated red hot, is used to burn a hole rather than to cut a hole.

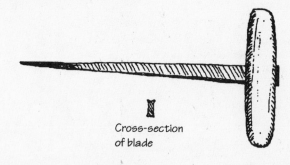

Cross-section of blade

• Auger, Carpenter's Nut

Before electric drills became common, this was the tool used by every woodworker for making larges holes. It is called a nut auger because a nut is used to fasten the wooden handle. A carpenter might have several sizes of nut augers, from ½ in. to 2 in., and the same handle could be used for all.

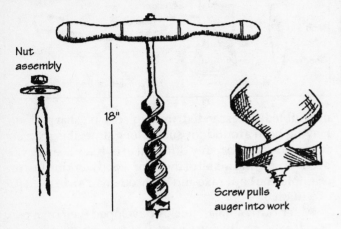

Nut assembly

18"

Screw pulls auger into work

• Auger, Raft (Ship Auger)
The raft (or ship) auger was made not for use on ships but for drilling holes in logs to be floated down rivers. The holes thus made were for the spikes and chains that kept the logs together in rafts. The augers were made long enough for a man to bore holes in the floating logs while standing on them, floating down river.

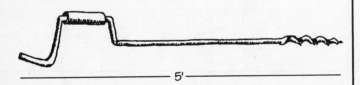

5'

• Auger, Tap
The small, one-handed tap auger not only drilled a hole, but because of its tapered sides, also made it bigger the further it penetrated. Thus this auger is in effect a *reamer* as well. Its chief use was for tapping barrels.

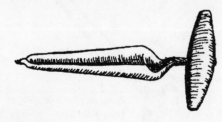

Auger Bit: see *Bit*.

Auger-bit File: see *File*.

Auger-bit Gauge: see *Bit Gauge*.

Awl
Basically, awls are no more than sharp points with wooden handles, used for making small holes. A variety of awls has been developed, however, all for distinct purposes.

• Awl, Belt
A belt awl is sometimes made with holes in the blade as shown, which fits it not only for piercing holes in belting but for pulling lacing through as well.

• Awl, Brad
This is the most common awl, often carried by carpenters and frequently spelled as one word, "bradawl." Its chief purpose is to make pilot holes for screws or starting holes for drilling.

• Awl, Carpet
Used for piercing holes in carpet, quality tools often have handles made of apple, maple, or even rosewood.

• Awl, Magazine Brad
An awl with a hollow handle containing a set, or "magazine," of different-size blades.

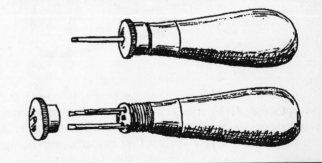

• Awl, Peg

The peg awl is used in caning chairs.

• Awl, Scratch

Used for general marking and measuring in woodwork, such as when laying out joints.

• Awl, Stabbing

A very small, all-purpose awl.

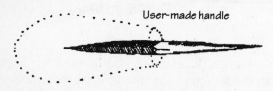

User-made handle

• Awl, Upholsterer's

Also known as a "cane-seating awl," the upholsterer's awl is a heavy awl used in the caning and rushing of chairs and stools. The blade is often driven entirely through the handle and then riveted at the end.

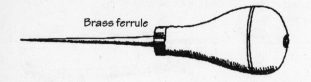

Brass ferrule

Axe

Along with the *hammer*, the axe is probably one of the oldest tools used by mankind, hitting and cutting being two very primal actions. As far as cutting goes, the *knife* and the axe are surely at least first cousins, both being

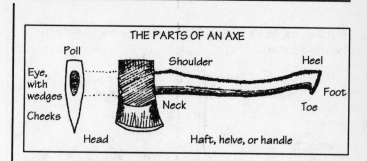

THE PARTS OF AN AXE

Poll
Eye, with wedges
Cheeks
Head
Shoulder
Heel
Foot
Toe
Neck
Haft, helve, or handle

undoubtedly descended from pieces of sharp stone. Having been around for such a long time, it is scarcely surprising that so many different axes have been developed—by no means all of them for woodworking (as, for example, the battle axe and the headsman's axe used for executions).

American colonists quickly developed their own particular axes, producing a variety of regional axes such as the Connecticut axe, the Michigan axe, the Yankee axe, the Western axe, the Puget Sound axe, and, most famous of all regional axes, the Hudson's Bay axe. A significant American contribution to axe design was the addition of

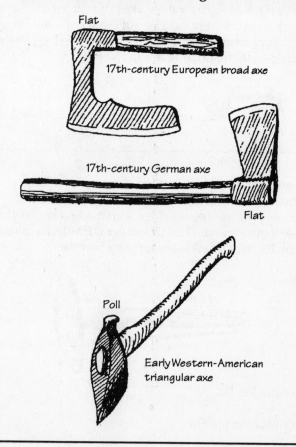

Flat

17th-century European broad axe

17th-century German axe

Flat

Poll

Early Western-American triangular axe

a heavy poll, giving more weight and thus more momentum to chopping. Earlier European axes were typically flat-backed.

• Axe, Broad

The broad axe, together with the *wood axe*, was the carpenter's axe of choice in the days of heavy wood-framing. Together with the *adze*, it was the tool used for hewing large beams. It is distinguished from other axes by having a proportionately larger head and smaller haft. Furthermore, only one side of the cutting edge is beveled, since the blade is commonly used flat against the work. This is also the reason why the haft (often handmade) is formed so that it emerges from the eye of the head at an angle, allowing the hand to clear the wood being trimmed.

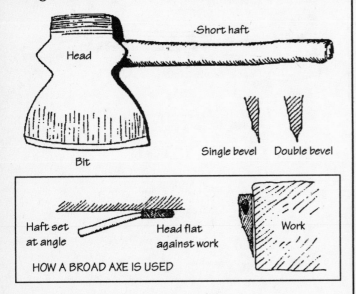

Short haft

Head

Bit

Single bevel   Double bevel

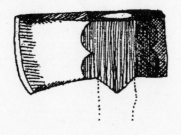

Haft set at angle   Head flat against work   Work

HOW A BROAD AXE IS USED

• Axe, Dock

The dock axe is a good example of the specialization of axes. Note the straight top of the head and the large poll to make chopping easier.

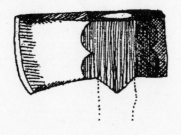

• Axe, Fire

The fire axe, or "fireman's axe," is intended mostly for use in factories, warehouses, hotels, and public buildings. Part of the head and sometimes the haft are painted red, and the axe is usually to be found hanging from, or sitting in, special brackets on the wall.

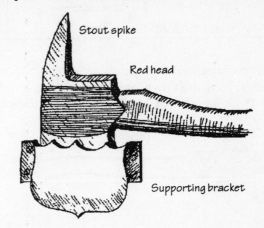

Stout spike

Red head

Supporting bracket

• Axe, Ice

This is a narrow-headed axe made with a solid-steel pick in place of a poll.

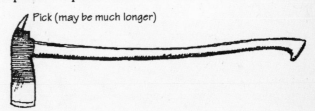

Pick (may be much longer)

• Axe, Mortise

A mortise axe is an old tool used for excavating the mortise part of mortise-and-tenon joints common in heavy timber-frame construction, and is often used in conjunction with the *mortise chisel*.

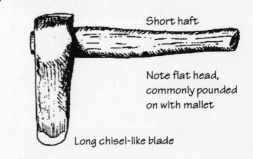

Short haft

Note flat head, commonly pounded on with mallet

Long chisel-like blade

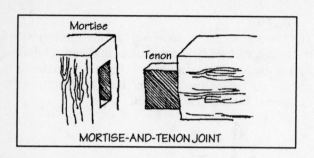

MORTISE-AND-TENON JOINT

• Axe, Ship Carpenter's

Shipbuilders have their own kinds of axes to accommodate the large amount of hand shaping and hewing

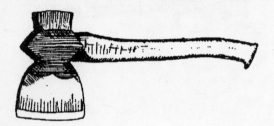

necessary in a wooden vessel. This axe may be beveled on both sides of its cutting edge.

• Axe, Wood

Wood axes are made in different weights ranging from 2 lb. to 10 lb., in 1-lb. increments. The most popular sizes are the 4-lb. hand axe used for light trimming, and the 7-lb. felling axe used for felling trees (6-lb. felling axes are preferred in Sweden). Haft shapes and sizes vary considerably; many craftsmen and woodsmen prefer to fit their own. Hafts are usually made of ash or hickory, which, while strong, are also flexible and absorb shock comfortably. A woodsman's axe may sometimes have two cutting edges: one for chopping and a second, finished with a finer taper to the edge, for trimming.

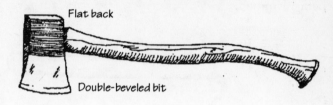

Flat back

Double-beveled bit

House Framing Tools, from *Orbis Sensualium Pictus,* by Johann Amos Comenius, 1685

Tools shown include: 4. felling axe, 7. wedge, 8. beetle,
10. broad axe, 13. crosscut saw, 15. jack, 17. chalk line.

The Joiner, from *Das Ständebuch* (The Book of Trades), by Jost Amman, 1568

Tools shown include: augers and chisels (in rack behind bench), hammer, jointer and smooth plane, dividers, square, plow plane, awl, and chisels (on bench), bow saw and axe (in foreground).

Back-bent Gouge: see *Chisel, Woodcarving.*

Backeroni: see *Chisel, Woodcarving, Box Tool.*

Back Saw: see *Saw.*

Ball-pein Hammer: see *Hammer, Machinist's.*

Bar Clamp: see *Clamp.*

Beading Plane: see *Plane, Moulding.*

Beading Tool
A beading tool is used for making moulding patterns in situations where *moulding planes* cannot be used, such as when working diagonally, across the work, or on irregularly shaped surfaces. (For a separate tool with the same name see *woodturning chisel*.)

• Beading Tool, Single-handed Beader
This tool is designed to cut beads, reeds, or flutes with one hand. While not as easy to control as the *universal hand beader*, it is invaluable in close quarters where there is little room for the larger tool.

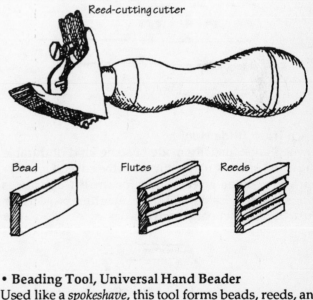

• Beading Tool, Universal Hand Beader
Used like a *spokeshave*, this tool forms beads, reeds, and flutes on straight and irregular surfaces. It can also be used for various kinds of light routing. It has a square fence for guiding it along straight work and an oval fence for use with curved work. Equipped with seven different

blades, it can make six sizes of beads, four sets of reeds, and two sizes of flutes. Using square-edged blades, it can also make two sizes of flat-bottomed grooves.

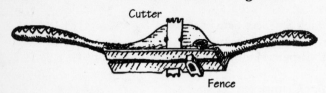

Cutter

Fence

Beetle: see *Maul*.

Beltmaker's Plane: see *Plane, Special-purpose*.

Bench Dog

Early bench dogs were invariably user-made out of any convenient piece of hardwood; contemporary bench dogs are made of metal. Both kinds are designed to be placed in any of the holes mortised in the front of a cabinetmaker's bench. Wooden ones are held by friction, while metal ones are held by a spring leaf attached to the side of the dog. Used in conjunction with *benchstops* or *vises* integral with the bench, bench dogs form an important part of the bench's holding system.

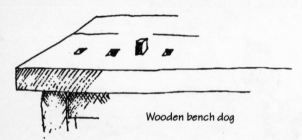

Wooden bench dog

Bench Hook (Side Hook)

A bench hook, usually made of some kind of durable hardwood such as maple, fits over the edge of a workbench to provide a secure rest for small work pieces being sawn or chiseled, and at the same time protects the bench itself from being marred.

Bench hook

Workbench

Benchstop

A benchstop is something fixed in the surface of a workbench to prevent wood that is being planed or otherwise worked on from being pushed across the surface. The simplest form of benchstop is a block of wood (similar in size to a *bench dog*) that may be raised or lowered in a hole mortised in the benchtop, and is usually secured in the desired position by a simple screw. Factory-made benchstops, like the one illustrated, are often entirely metallic. Although benchstops made of cast iron were once common, contemporary models are made with softer metal to avoid damaging tool edges.

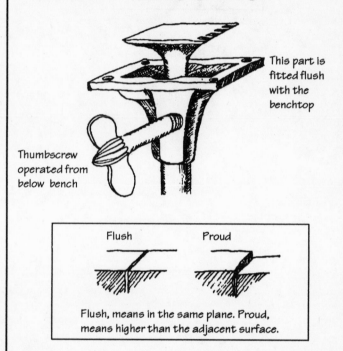

This part is fitted flush with the benchtop

Thumbscrew operated from below bench

Flush   Proud

Flush, means in the same plane. Proud, means higher than the adjacent surface.

Bevel

A bevel is a tool for checking and marking angles. The earliest bevels consisted simply of two pieces of wood pinned together at one end.

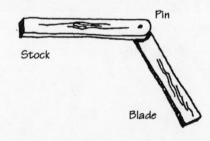

Pin

Stock

Blade

• Bevel, Combination

The addition of an extra, slotted stock to this bevel makes possible the measuring and laying out of any desired angle flat upon the work. Made from metal, this is primarily a machinist's tool.

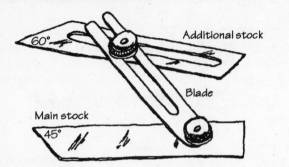

Additional stock

60°

Blade

Main stock

45°

• Bevel, Improved

Another variety of a machinist's bevel, the improved bevel has the advantage that both the stock and the blade are slotted, thus making possible many adjustments that could not be made with a common bevel.

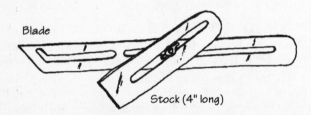

Blade

Stock (4" long)

• Bevel, Protractor (Universal Bevel Protractor)

The bevel protractor is a very sophisticated bevel that also includes the properties of a protractor. One side is completely flat, making it possible to use this tool for drafting as well as for laying out work.

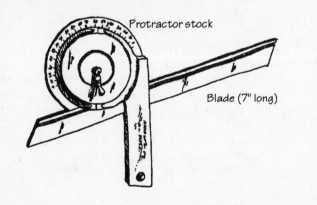

Protractor stock

Blade (7" long)

• Bevel, Sliding-T

A common woodworking bevel, the sliding-T bevel has a thick stock and a thin, slotted blade that allows it to be pivoted from the end as well as from any point along the slot. The metal-stock variety uses a screw in the base of the stock to secure the angle of the blade; the more common wooden-stock variety is secured directly at the pivot point.

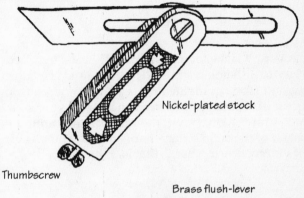

Steel blade

Nickel-plated stock

Thumbscrew

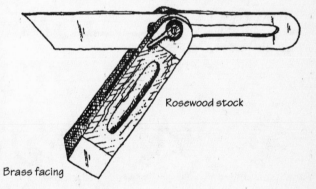

Brass flush-lever

Rosewood stock

Brass facing

• Bevel, Universal

The offset in the heel of the universal bevel's blade makes this tool especially valuable for machinists when turning bevel-gear blanks or similar work.

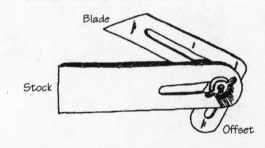

Blade

Stock

Offset

Bit

A bit is that part of a *drill* or *brace* (tools used for making holes) which actually penetrates the work. It cannot be used on its own—it always needs a drill or brace to turn it. Since the same drill or brace can hold many different sizes it is only necessary to have one drill or brace and a set of bits to be able to bore an entire range of holes.

It is the woodworker who calls this tool a bit; the machinist calls it a "drill." To avoid confusion, the machinist refers to the tool that holds his "drill" as a *hand drill* or *breast drill*, as the case may be.

• Bit, Auger

The auger bit is the bit most commonly used by woodworkers to drill holes. It is used in a *brace* and is made in a range of sizes, graduated in 1/16–in. increments, from 3/16 in. to 1 in. or slightly more. The bit size is generally stamped on the bit's tang, the number indicating the diameter of the bit in sixteenths of an inch. Thus, *8* indicates a bit measuring 8/16 in., that is, 1/2 in.

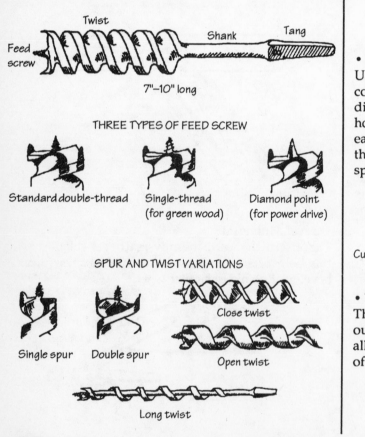

Twist
Feed screw
Shank
Tang

7"–10" long

THREE TYPES OF FEED SCREW

Standard double-thread | Single-thread (for green wood) | Diamond point (for power drive)

SPUR AND TWIST VARIATIONS

Single spur | Double spur

Close twist

Open twist

Long twist

• Bit, Bell-hanger's Twist

Bell-hanger's twist bits are exceptionally long, ranging from 1 ft. to 3 ft. in length. Unlike *auger bits* they are sized in 1/32–in. increments.

• Bit, Bright Spoon

The bright spoon bit is a simple, gouge-shaped bit used mostly for boring holes in wood.

• Bit, Caster

The caster bit is a short bit that can cut through wood and nails without injury to itself. It may also be called a "single-twist bit" since, unlike regular *twist bits*, it is made with only one cutting edge.

• Bit, Center (Br. Centre Bit)

Unlike *auger bits*, center bits have no twist to them, and consequently their cutting action is only in a downward direction. For this reason they are used mainly to start holes or to make shallow borings. They were among the earliest bits to be used with a *brace*, and yet are basically the same design as the contemporary speed-bore or spade bits used with electric drills.

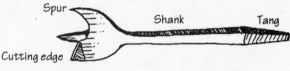

Spur
Shank
Tang
Cutting edge

• Bit, Countersink Gimlet

This bit bores a hole for a screw. The bit has the same outline as a woodscrew, and the hole it bores thereby allows the screw to be screwed in flush with the surface of the work.

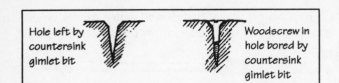

Hole left by countersink gimlet bit

Woodscrew in hole bored by countersink gimlet bit

• Bit, Dowel

The dowel bit is also known as a "short auger bit"— which is exactly what it is. Its main use is for boring the holes into which dowels are to be inserted.

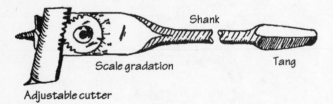

Hole made by dowel bit

Dowels

• Bit, Expansive (Br. Expanding Bit)

An expansive bit (generally made in two slightly over-lapping sizes, each with two adjustable cutters that make possible a range of variously sized holes) is, in essence an adjustable *auger bit*. The larger expansive bit can bore holes up to 4 in. in diameter.

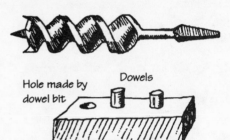

Shank

Tang

Scale gradation

Adjustable cutter

• Bit, Forstner

Originally spelled "foerstner," and sometimes seen as "fostner," forstner bits are used to bore holes that do not penetrate the work completely and where a flat-bottomed hole is needed. (The spurs or screw of a regular *auger bit* might penetrate the work even if the bit were stopped short of the bottom.) They are also preferable when boring in end-grain, thin wood, or other places where an auger bit might cause splitting.

Cutting edge

One disadvantage of the forstner bit is that, having no feed screw, it is often necessary to scribe a circle the size of the bit on the work before the bit may be centered.

• Bit, German Gimlet

This is a large *gimlet* for use with a *drill* or *brace*. Sized from $\frac{1}{16}$ in. to $\frac{3}{8}$ in., in increments of $\frac{1}{32}$ in., it is found up to 12 in. long. The gimlet bit makes a hole by actually removing wood. The point centers the bit on the hole, and the outer edge of the spiral does the cutting. The chips thus cut are collected in and carried up the flute as the bit is turned.

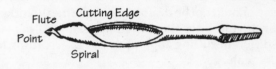

Flute

Cutting Edge

Point

Spiral

• Bit, Plug Cutter

A plug cutter cuts a circle, leaving a core or plug that may then be removed and used to plug another hole, such as the hole left by a recessed screw.

Early, open type

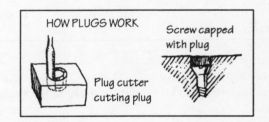

HOW PLUGS WORK

Screw capped with plug

Plug cutter cutting plug

• Bit, Ship Auger

Ship auger bits are much longer than regular *auger bits*. They have no spurs, a very open twist, and may be used with or without feed screws. Also known as "spiral augers," they are often used with a transverse wooden handle in the same way as a *carpenter's nut auger*.

Open spiral

• Bit, Taper

Taper bits are relatively small bits, similar to *tap augers*, and are used in *hand drills* for reaming out small holes such as barrel bung-holes.

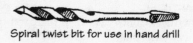

• Bit, Twist

The twist bit is used to bore holes in wood for screws, nails, and bolts. Although intended primarily for wood, good quality wood twist bits are often warranted to cut through nails without injury to the bit. The twist bit, however, should not be confused with the metalworker's twist drill which, despite similar appearances, is made with a different cutting angle and is sized according to a different scale.

Spiral twist bit for use in hand drill

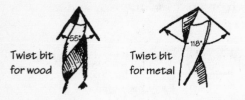

Twist bit for wood

Twist bit for metal

Bit Brace: see *Brace*.

Bit Gauge (Auger Bit Gauge)

This is a stop which may be fixed at any point on an *auger bit* to ensure that the bit penetrates only to the required depth.

Bit Holder (Extension Bit Holder)

Early bit holders were made with a simple socket for the tang of the bit, which was secured by a wingnut. But the effect was the same as when using the more modern variety illustrated here—namely, to allow the user to bore through walls, floors, and other places where an ordinary bit would be too short to complete the hole.

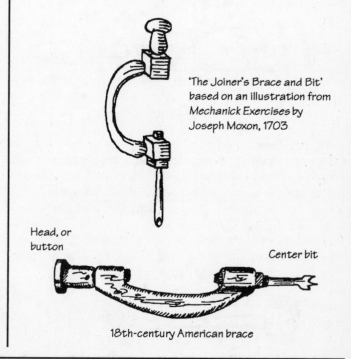

Block Plane: see *Plane, Block*.

Bolster Chisel: see *Chisel, Cold*.

Bow Saw: see *Saw*.

Bowl Gouge: see *Chisel, Wood, Gouge*.

Box Chisel: see *Chisel, Wood, Special-purpose*.

Box Scraper: see *Scraper*.

Brace

The brace, also known as a "bit brace," and formerly as a "bitstock," is basically a crank for holding and turning the various *bits* used for boring holes. There are two basic types: the plain brace and the ratchet brace. The brace is thought to have been developed in the fifteenth century, for there are no earlier references to it.

'The Joiner's Brace and Bit' based on an illustration from *Mechanick Exercises* by Joseph Moxon, 1703

Head, or button

Center bit

18th-century American brace

Until the nineteenth century, most braces were plain, one-piece, wooden braces. Thereafter many braces became very sophisticated; metal parts were introduced, the head and handle were made movable, and the chucks for holding the bits were improved. The tool was brought to a state of near perfection with the introduction of the type illustrated, known as the "Ultimatum."

19th-century 'frame brace,' made of ebony, brass, and ivory

• Brace, Corner Bit

By the early twentieth century, most braces were being made mainly of metal, with ball-bearing joints in the head and handle and ratchet chucks for holding the *bits*. Also at this time many specialty braces were developed, such as the corner bit brace, which enables the worker to use the tool in an otherwise difficult or impossible situation (such as a corner).

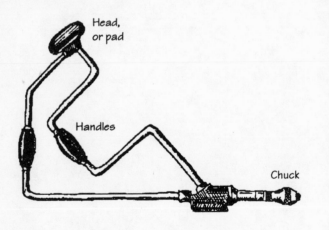

• Brace, Combination Corner

One of the more exotic braces of the early twentieth century, this tool is a combination *ratchet brace*, plain brace, and *corner brace*. It holds any size *bit*, as well as small, round-shank *drills*. It has positive drive, no lost motion, and will drive a bit true at any angle. Made with

a 10–in. sweep and a ball-bearing head, it was often highly polished and nickel-plated.

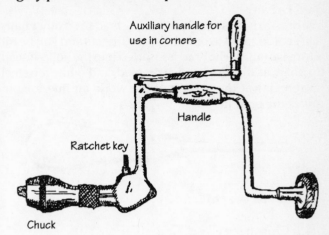

• Brace, Ratchet

The ratchet brace was the standard handtool carried by most carpenters until the advent of the cordless electric drill. The ratchet brace is indispensable for use in situations where a full turn of the handle is impossible, for the ratchet permits the brace to be turned in the opposite direction from that required to work the *bit*, without causing this to back out. The ratchet collar controls the direction in which the ratcheting action works.

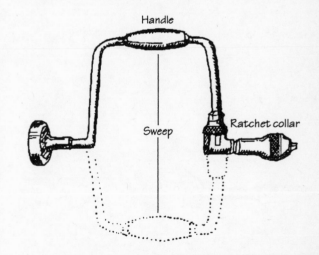

"Sweep" is the diameter of the arc made by the handle as it is revolved. Braces are commonly sized by their sweeps, for example, an 8 in. brace indicates one with a sweep of 8 in., not one that is 8 in. long.

Brad Awl: see *Awl*.

Brad Driver
The brad driver is a tool for driving brads and other nails that are too small to be held by the fingers and still be hit with a hammer. The brad is inserted into the hollow steel tip; then, as the handle is depressed, and while the steel tip remains at the surface of the work, an interior pin drives the brad in.

Hollow steel tip

Breast Drill: see *Drill*.

Brick Chisel: see *Chisel, Cold*.

Broad Axe: see *Axe*.

Buck Saw: see *Saw*.

Bull-nose Rabbet Plane: see *Plane, Special-purpose*.

Bull's-eye Level: see *Level*.

Burnisher: see *Scraper Burnisher*.

Butt Chisel: see *Chisel, Wood, Bench*.

Butt Gauge: see *Gauge*.

Butt Marker
A butt marker is held against the edge of a door, window, or jamb—in fact, wherever a hinge (or "butt," as certain hinges are known) is to be fitted—and then struck hard with a *hammer* or *mallet*. When it is removed, it will be seen that its sharp edges will have bitten into the wood, thereby marking the area to be mortised to receive the hinge.

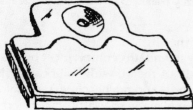

The Turner (top), **The Boxmaker** (bottom), from *Orbis Sensualium Pictus*, by Johann Amos Comenius, 1685

The Wagonwright, from *Das Ständebuch* (The Book of Trades), by Jost Amman, 1568

Tools shown include: wheelwright's reamer, axe, and mallet.

C

Cabinetmaker's Block Plane: see *Plane, Block.*

Cabinetmaker's Scraper: see *Scraper.*

Cabinet Rasp: see *Rasp.*

Cabinet Scraper: see *Scraper.*

Calipers (Br. Callipers)

A pair of calipers is an instrument with bowed legs for measuring the diameter of convex bodies, originally bullets or cannonballs, which were then said to be of such-and-such "caliber." Hence, derivatively, phrases like "a heavy-caliber gun." It is interesting to note that from the beginning, the two words have been spelled differently (in British English they are spelled "calliper" and "calibre"). As with many other tools and instruments, specialization has occurred, and a pair of calipers may also be an instrument with straight legs and points turned outwards for measuring the bore of tubes and other hollow objects. To avoid confusion, these are known as *inside calipers*; the original variety now being called *outside calipers*.

• Calipers, Double
This instrument may be used not only as *inside calipers* and *outside calipers*, but as *dividers* as well.

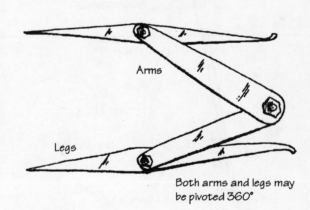

Arms

Legs

Both arms and legs may
be pivoted 360°

• Calipers, Inside
These calipers are used for measuring inside diameters. They are made in a variety of sizes, enabling them to measure from ¼ in. to as much as 2 in. Two types are shown: plain and spring. Spring calipers permit a finer

and surer adjustment, and are often known as "bow spring calipers."

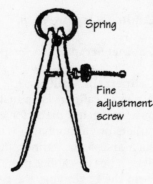

Plain inside calipers   Spring inside calipers

Spring

Fine adjustment screw

• Calipers, Keyhole

Keyhole calipers are an example of one of the many specialized uses to which calipers have been adapted.

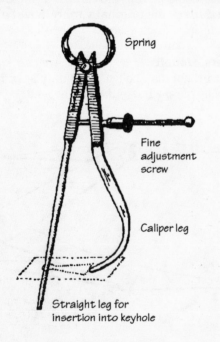

Spring

Fine adjustment screw

Caliper leg

Straight leg for insertion into keyhole

• Calipers, Outside

Outside calipers, as mentioned above under *calipers*, were the original instrument for measuring the diameter of convex bodies such as cannonballs or pipes. They may be made as either of the two types shown under *inside*

calipers (plain or spring), or "winged," as shown here—as, indeed, may *inside calipers* themselves.

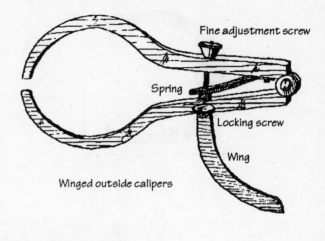

Fine adjustment screw

Spring

Locking screw

Wing

Winged outside calipers

Caliper Rule: see *Rule*.

Cape Chisel: see *Chisel, Cold*.

Carpet Stretcher

This tool has little to do with carpentry or woodworking, but is included as a representative of a large group of furnishers' and upholsterers' tools that includes tools originally designed for woodworking but are now adapted for other uses as well (such as the carpet vise, the carpet-sewer's clamp, webbing pliers, and the *carpet awl*). The carpet stretcher illustrated here is a particularly fine tool, being made of forged steel with extra strong teeth, a jointed hickory handle with a lignum vitæ head, and an 8–in. long, heavy brass ferrule. The entire tool measures 32½ in. from end to end.

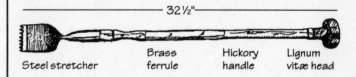

32½"

Steel stretcher   Brass ferrule   Hickory handle   Lignum vitæ head

Carriagemaker's Iron T-plane: see *Plane, Special-purpose, Rabbet*.

Carriagemaker's Panel Router

This tool is closely related to the *beading tool*, the *moulding plane*, and the *spokeshave* in that it combines elements of all three. Yet it is not properly a beading tool, since it doesn't cut beads (which are in relief from the work), but grooves (which are by definition recessed); neither is it a moulding plane, since it is meant not for flat but for curved surfaces; nor is it a spokeshave, since it is used neither to shave nor for spokes. What it does do is to cut varying-sized grooves in irregular and curved surfaces.

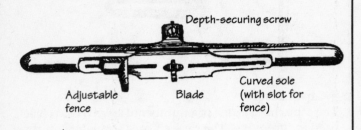

Depth-securing screw

Adjustable fence · Blade · Curved sole (with slot for fence)

Carver's Screw

The carver's screw is a stout screw about 12 in. long, used to secure the work to the bench. The screw, which has a distinctive head used to tighten the screw and clamp the work to the bench, is inserted through a hole bored in the benchtop into the bottom of the workpiece.

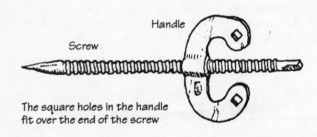

Handle

Screw

The square holes in the handle fit over the end of the screw

Cat's Paw: see *Nail Claw.*

Cavetto Plane: see *Plane, Moulding.*

C-clamp: see *Clamp.*

Center-bead Beading Plane: see *Plane, Moulding.*

Center Bit: see *Bit.*

Center Punch (Br. Centre Punch)

A center punch is a small, steel tool, much used by metalworkers, but also frequently employed by woodworkers, for making small holes or indentations in wood

or metal so that screws or *drills* may be started without wandering.

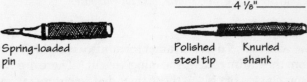

4 ⅛"

Spring-loaded pin · Polished steel tip · Knurled shank

Automatic center punch · Regular center punch

Chalk Line

A chalk line is used for marking a straight line longer than the longest available *straight edge.* The line, which is coated with chalk, is stretched taut, close to the work, and then plucked, or "snapped," thereby depositing a straight line of chalk on the work.

• Chalk Line, Cased

Present-day chalk lines usually consist of a line in a metal case filled with powdered chalk. The line may be withdrawn and reeled back as desired. It is generally provided with a metal hook or ring at the free end, preventing it from being reeled back in completely, as well as allowing it to be hooked over one end of the work—making it possible for one person to snap a line unaided.

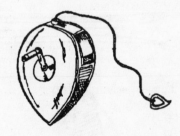

• Chalk Line, Reel and Chalk

Until the invention of the *cased chalk line,* the term *chalk line* implied the use of a reel and chalk. The line was stored on a reel and rubbed against a block of chalk—usually sold as a sphere or half-sphere—before use.

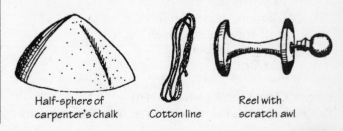

Half-sphere of carpenter's chalk · Cotton line · Reel with scratch awl

Chamfer Knife: see *Drawknife.*

Chamfer Plane: see *Plane, Special-purpose.*

Chisel

The word *chisel* derives rather tortuously via Old French from the Latin: *cædere,* meaning "to cut." There are two main classes of chisel: those intended to cut metal and stone, listed below under **Chisel, Cold,** and those intended to cut wood, which are listed under **Chisel, Wood.**

Chisel, Cold

Cold chisels are intended to cut metal or stone, usually by being struck with a *hammer* or heavy *mallet.* Apart from the *star drill,* whose shank is sometimes fitted with a plastic sleeve to make holding the tool more comfortable when it is being struck, cold chisels, unlike wood chisels, are handleless. The most common variety is known simply as a "cold chisel," and is the sort typically carried by carpenters and other mechanics to knock out a hole in a wall of stone or brick for the insertion of a wedge, or to receive the end of a piece of timber. Metal-workers use the cold chisel for simple cutting operations.

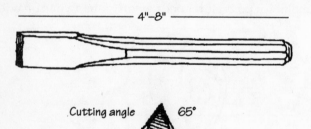

The following varieties of cold chisel are listed alphabetically:

    Brick Chisel
    Cape Chisel
    Diamond-nose Chisel
    Half-roundnose Chisel
    Plugging Chisel
    Star Drill

• Brick Chisel (Br. Bolster Chisel)

The brick chisel is the widest of the cold chisels, measur-

ing on average 3 in. to 3 ½ in. As the name implies, it is used for cutting bricks or masonry to size.

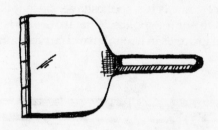

• Cape Chisel

The cape chisel is one of a number of special cold chisels used for grooving. The cape chisel's specialty is cutting keyways—slots holding "keys" (small pieces used to keep adjacent parts aligned).

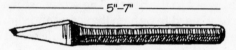

• Diamond-nose Chisel

Also known as a "diamond-point chisel," this variety of cold chisel is used for cutting grooves in metal.

Diameter	3/8"	1/2"	5/8"	3/4"
Length	5"	5 3/4"	6 1/4"	7"

TYPICAL DIMENSIONS

• Half-roundnose Chisel

The half-roundnose chisel is an adaptation of the *round-nose chisel.* It is used for making round grooves in metal.

• Plugging Chisel

The plugging chisel is one example of the large variety of of chisels used for purposes other than cutting metal, stone, or wood. Like the boring chisel, tamping chisel, and other offset cold chisels, it is used for jobs such as setting oakum between planks or tamping lead wool into soil-pipe joints.

• Round-nose Chisel

A round-nose cold chisel is the standard tool for cutting round grooves in metal.

• Star Drill

The star drill is a kind of cold chisel in that it is made of steel, is hit with a *hammer*, and is used for working with stone or masonry. It is called a "drill" since its main job is to make a hole, although it is used more like a chisel than an actual *drill*.

Alternative pattern, sometimes called a 'pipe drill'

Chisel, Wood

A wood chisel is usually understood to be a tool consisting of a metal blade, sharpened into a cutting edge at one end and handled at the other, intended to cut wood. Few woodworking tools have become as diversified as the wood chisel, and therefore, in order to maintain as much

perspective as possible, they are here divided into five main groups as follows:

    1. **Bench Chisels**
    2. **Special-purpose Chisels**
    3. **Gouges**
    4. **Woodturning Chisels**
    5. **Woodcarving Chisels**

None of these classifications is totally exclusive since some tools could legitimately be included in more than one group; they are merely a convenience in describing so many chisels.

1. Bench Chisels

This group comprises, in the following order of relative importance:

    Firmer Chisels
    Paring Chisels
    Butt Chisel
    Plastic-handled Wood Chisel
    Mortise Chisel
    Corner Chisel
    Framing Chisel
    Notching Chisel

• Firmer Chisel

The firmer chisel gets its name from the Old French chisel known as the *fermoir*, which was used for making mortises (*see box on page 6*). It is now the general-purpose chisel used by carpenters on site and by joiners, furnituremakers and cabinetmakers at the bench for a variety of jobs.

• Firmer Chisel, Bevel-edge

This is a better quality *firmer chisel* made with beveled sides instead of square sides. The bevels allow the cutting edge of the chisel to penetrate more closely into corners.

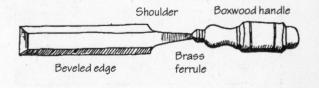

Shoulder   Boxwood handle
Beveled edge   Brass ferrule

• Firmer Chisel, Socket

The socket firmer chisel is stronger than the regular *firmer chisel*. Its socket construction allows the chisel to withstand heavier blows and be used for harder work.

Socket

• Firmer Chisel, Bevel-edge Socket

This is the top-of-the-line *firmer chisel*, incorporating all the best features of chisel construction: a socketed and beveled blade often incorporating a quality hardwood or fruitwood handle, fitted with a leather tip (to prevent the end of the handle from fraying and splitting when hit repeatedly with a mallet).

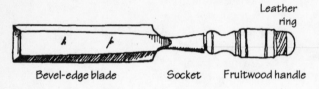

Bevel-edge blade   Socket   Fruitwood handle

Leather ring

• Paring Chisel

A paring chisel is a generally lighter-duty tool than a *firmer chisel*, and is used mainly for shaping and final trimming. Since it is intended for lighter work than the firmer chisel, it is generally manipulated by the hand alone. For this reason a socket is not normally necessary, a tanged construction being sufficient. Similarly, the bevel of its cutting edge is much shallower, since sharpness is more important than strength.

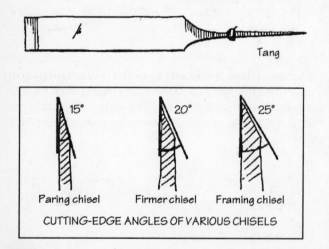

Tang

15°   20°   25°

Paring chisel   Firmer chisel   Framing chisel

CUTTING-EDGE ANGLES OF VARIOUS CHISELS

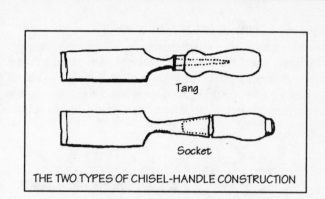

Tang

Socket

THE TWO TYPES OF CHISEL-HANDLE CONSTRUCTION

• Paring Chisel, Bevel-edge

Bevel-edge paring chisels are the rule, although some workers prefer regular paring chisels with straight sides.

Width: ¹⁄₁₆", ¹⁄₈", ³⁄₁₆", ¹⁄₄", ⁵⁄₁₆", **³⁄₈"**, ⁷⁄₁₆", **¹⁄₂"**, ⁵⁄₈", **³⁄₄"**, ⁷⁄₈", **1"**, 1 ¹⁄₄", 1 ¹⁄₂", 1 ³⁄₄", **2"**

Firmer and paring chisels are made in all the above widths; those in bold face are the most common.
CUTTING-EDGE WIDTHS

• Paring Chisel, Patternmaker's Chisel

An extra-long, bevel-edge paring chisel made with a cranked neck is known as a patternmaker's chisel, although its use is far from restricted to patternmakers alone.

Cranked neck

• Paring Chisel, Socket

The blades of the best chisels used to be made of selected steel, oil-tempered and carefully tested. The blade widens slightly towards the cutting edge. The ferrule and blades of socket chisels are carefully welded together to form one piece.

Blade   Socket (ferrule)

• Paring Chisel, Bevel-edge Socket

Beveled edges are preferable to square edges because they tend to drive the tool forward and have greater clearance, especially in corners. It should be pointed out that the backs of most bench chisels are perfectly flat, and those with blades of widths up to about 1 in. start life about 8 in. long—repeated sharpenings, of course, reducing their length.

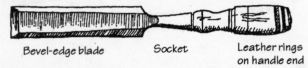

Bevel-edge blade   Socket   Leather rings on handle end

• Butt Chisel

The butt chisel is distinguished by having a blade only 2 ½ in. to 3 ¼ in. long when new. This makes it particularly well suited for installing small hardware—such as butts (hinges)—where little power but a wide cut is required. A chisel with a blade from 4 in. to 5 in. long is properly called a "pocket chisel." Its use is practically the same; its name derives from the ease with which it may be kept in a pocket.

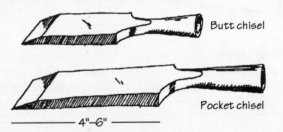

Butt chisel

Pocket chisel

4"–6"

• Plastic-handled Wood Chisel

This is the contemporary jack-of-all-trades chisel, commonly used by carpenters and other workers on construction sites. It is made in a variety of widths, but the blade is generally only 3 in. to 3 ½ in. long. The metal-capped plastic handle is cast and moulded in place, and is consequently very sturdy, making this an excellent all-round, heavy-duty chisel, capable of being hit with a *hammer*, although the shortness of the blade prevents a good job of paring or deep mortising.

3" blade   Plastic handle with steel tip

• Mortise Chisel

The mortise chisel is made much stronger than other chisels since it must withstand many heavy blows. It has a much squarer cross-section than other bench chisels in order to resist the levering necessary when cutting mortises (*see box on page 6*). There are several varieties, of which the longer, socketed version, originally intended for making small mortises, is more common. The stouter, oval-handled version, often provided with a leather washer between the bolster and the handle to absorb shock, was intended for heavier use by joiners.

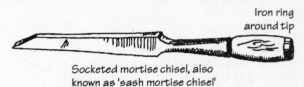

Iron ring around tip

Socketed mortise chisel, also known as 'sash mortise chisel'

• Corner Chisel

The corner chisel is another chisel employed in the cutting of large mortises—especially those forming part of heavy timbering (*see box on page 6*). It is used to clean out the corners of the mortise. Heavier mortise chisels were originally used without handles; a bent and battered socket end (sometimes called the "cuff") indicates that a chisel was used this way.

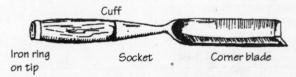

Cuff

Iron ring on tip   Socket   Corner blade

• Framing Chisel

The framing chisel is a heavier version of the *firmer chisel* (*see diagram of cutting-edge angles under Paring Chisel*) and was originally used when making the tenons for the mortises cut by the firmer chisel. Being intended for harder work than the firmer chisel, it is often fitted with an iron band on the handle to prevent it splitting.

Framing chisels may be completely flat—back and front—or they may have a slight curve on the bevel side like the one illustrated.

Note rounded profile of bevel side

• Notching Chisel

Notching chisels are thin but strong and stiff chisels made with no bevel, used by stair builders. They are made in several sizes, from 1 in. to 2 in. wide, and from 4 in. to 5 in. long.

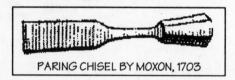

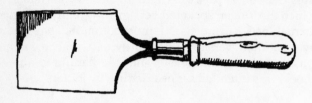

PARING CHISEL BY MOXON, 1703

2. Special-purpose Chisels

This is a group of single-purpose chisels, generally much larger than those of the previous group. They are listed in alphabetical order as follows:

    Box Chisel
    Deck Chisel
    Floor Chisel
    Peeling Chisel
    Plumber's Chisel
    Ripping Chisel
    Slick

• Box Chisel

Box chisels are long, steel bars—they may be round, flat, or even octagonal in cross-section—with a cleft cutting edge, like the claw of a *nail hammer*. They are used in the opening and destruction of boxes and crates.

16"

• Deck Chisel

A deck chisel ranks in size between the *framing chisel* (described in the previous group) and the *slick* (the last chisel described in this group). Much used in wooden ship and boat construction, the beveled side of the blade is always rounded. It is found in two main sizes: 1 ¼ in. and 1 ½ in. wide, both sizes being approximately 5 in. to 6 in. long.

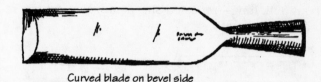

Curved blade on bevel side

• Floor Chisel

This is a long (18 in.) chisel made with no handle, intended to be hit with a *hammer*. The blade is made thin but wide to cut flooring and other long boards. Despite its appearance, it is not a *ripping chisel*, and is not intended for rough usage.

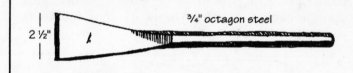

2 ½"   ¾" octagon steel

• Peeling Chisel

The peeling chisel, together with the *adze*, is the tool used for taking the bark off logs in order to hasten their drying so they may be used for construction.

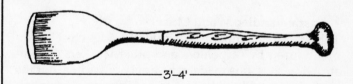

3'-4'

• Plumber's Chisel

This is a chisel used by plumbers when it is necessary to cut away parts of the wood of a house in order to install pipes. Similar to the chisel used by electricians, it is basically a heavy-duty, handleless chisel about 11 in. long and 1 in. wide.

• Ripping Chisel

A ripping chisel is intended for use in rough work, such as ripping down partitions. It should not be confused with a *ripping bar*.

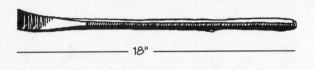

18"

• Slick

The slick is really a giant *paring chisel*, intended to be used with two hands. The end of the handle is often nestled against the shoulder, for which purpose the end is frequently found formed into a comfortable mushroom shape. It functions much like a plane in that it can be used to remove much more wood at a single pass than a small chisel. Some slicks are constructed so that the handle is slightly offset from the blade, thereby providing better clearance for the hands.

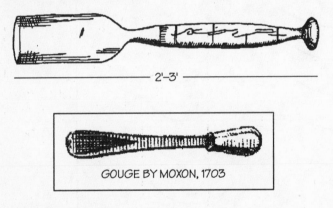

2'–3'

GOUGE BY MOXON, 1703

3. Gouges

A gouge is a concave chisel for cutting rounded grooves or holes in wood. There are four main classes of gouges: carpenter's gouges, patternmaker's gouges, turner's gouges, and woodcarver's gouges. In this section only carpenter's and patternmaker's gouges are described; turner's gouges are dealt with at **4. Woodturning Chisels**, and woodcarver's gouges are described at **5. Woodcarving Chisels**. The gouges here described are:

> Bowl Gouge
> Firmer Gouge
> Socket Firmer Gouge
> Paring Gouge
> Patternmaker's Gouge
> Plumber's Gouge

• Bowl Gouge

This is a short, bowl-shaped gouge, used for general scooping and gouging work.

Curved blade and cutting edge

• Firmer Gouge

Carpenter's gouges have blades with parallel sides, whereas woodcarver's gouges often become wider towards their cutting edge. Firmer gouges are generally beveled on the outside (such gouges being known as "out-cannel gouges"), although they may also be ground on the inside (such gouges being known as "in-cannel gouges"). They are made in a range of widths from ⅛ in. to 2 in., all sizes generally being made with a medium sweep (*see box on Sweeps*).

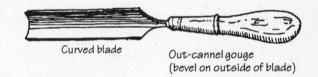

Curved blade

Out-cannel gouge
(bevel on outside of blade)

• Socket Firmer Gouge

Since the firmer gouge is meant for heavier work than the *paring gouge* (just like the *firmer chisel* and the *paring chisel*), more expensive tools are invariably constructed with a socket rather than a tang, the better to withstand heavy blows from a *mallet*.

Outside bevel

Socket

• Paring Gouge

Paring gouges are made with an inside bevel, and are designed for lighter, finer work than *firmer gouges*. They are made with regular, middle, and flat sweeps (*see box on Sweeps*) and include a few extra sizes not shown, such as ³⁄₁₆ in. ⁵⁄₁₆ in., and ⁷⁄₁₆ in. They are invariably made with tangs rather than sockets.

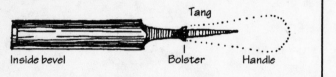

Tang — Bolster — Handle — Inside bevel

• Patternmaker's Gouge

Like the *patternmaker's chisel*, the patternmaker's gouge is longer than usual and made with a cranked neck. It is considered useful for the finest work, and is used by cabinetmakers and furnituremakers.

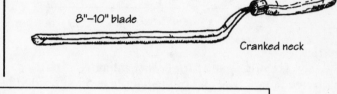

8"–10" blade — Cranked neck

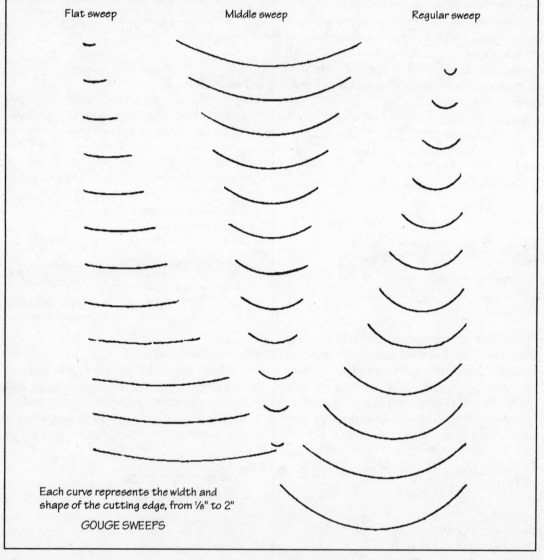

Flat sweep   Middle sweep   Regular sweep

Each curve represents the width and shape of the cutting edge, from ⅛" to 2"

GOUGE SWEEPS

• Plumber's Gouge

The plumber's gouge is the gouge counterpart of the *plumber's chisel*. It is a strong, handleless gouge intended for rough, heavy work.

11"

4. Woodturning Chisels

Woodturning chisels (often called simply "turning chisels") are those cutting or scraping tools used by the woodturner (usually referred to as a "turner") in shaping work (such as chair legs and other cylindrical objects) being turned on a lathe. The term *turning chisels* includes not only chisels but also gouges, as well as tools that are strictly neither. By convention they are all referred to as *chisels* in distinction to various other, non-cutting or non-scraping tools used in lathe work. The main difference between joiners' chisels and turners' chisels lies in size and strength. Turning chisels tend to have shorter, stouter blades but much longer handles. The main types are as follows:

     Turning Chisel
     Turning Gouge
     Beading Tool
     Parting Tool

• Turning Chisel

Turning chisels that are actual chisels are made with a tang, and their cutting edge may have one of three different profiles: square, round, or skew. All three shapes are made in a range of sizes from ⅛ in. to 2 in.

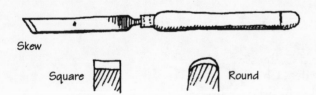
Skew
Square
Round

• Turning Gouge

Like *turning chisels*, turning gouges have short blades averaging only 6 in. to 7 in., but long handles. The cutting

13"

edge is beveled on the outside, and the sides of the blade are parallel.

• Beading Tool

This should properly be called a *woodturning beading chisel* to avoid confusion with the *beading tool* used to make moulding. In reality it is a form of reverse *gouge* used for making beads on turned work.

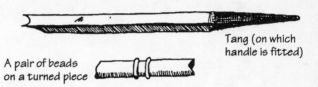

A pair of beads
on a turned piece

Tang (on which
handle is fitted)

• Parting Tool

A parting tool is a narrow-bladed turning chisel used for cutting V-shaped grooves and recesses in turned work. Parting tools are also sometimes called "V-tools," because the blade itself has a V-shaped cross-section. Parting tools, like *gouges*, are made in a variety of sizes and cross-sections, some "V"s being steeper than others.

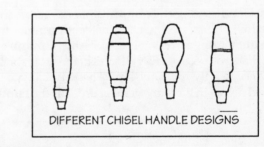
Unhandled tanged parting tool

DIFFERENT CHISEL HANDLE DESIGNS

5. Woodcarving Chisels

Woodcarving chisels comprise a very large and varied group of tools, not all of which are actual *chisels*, although the woodcarver typically refers to them as such. They range in size from ⅛ in. to 1 in., with a few smaller and larger exceptions. Since many of the better quality tools come from continental Europe, they are frequently sized in millimeters.

    Woodcarving chisels are usually of tang construction, and most woodcarvers prefer octagon-shaped handles for a better grip and also to prevent the tools from inadvertently rolling around. There is, however, an almost limitless variety of handle shapes. Many carvers advocate different handles as an aid to quick tool identification, especially when many are spread out on the workbench. Although woodcarving chisels do exist in a much shorter form than the average 9–in. tool, these are (apart from those intended for the block-cutter), inferior things meant only for the hobbyist and not for the professional woodcarver.

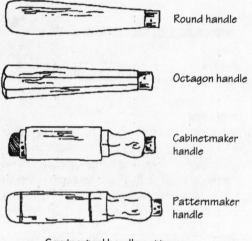

Round handle

Octagon handle

Cabinetmaker handle

Patternmaker handle

Carving-tool handle patterns

Most woodcarving chisels are numbered according to the "Sheffield Illustrated List," a nineteenth-century trade catalog that categorized these tools by a numbering system that grouped together all tools of the same shape— but not necessarily the same size. The various types are described below in the same general order as the number-ing system, their Sheffield numbers included in parentheses.

• **Firmer Chisel (1)**
The woodcarver's firmer chisel differs from the carpenter's *firmer chisel* by having a cutting edge that is beveled on both sides and at a much smaller angle—usually 12° to 15°, compared to 20° for the carpenter's tool.

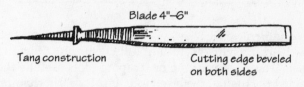

Blade 4"–6"

Tang construction

Cutting edge beveled on both sides

• **Skew Chisel (2)**
Like the firmer chisel, the skew chisel has a cutting edge beveled on both sides. Therefore it may constitute either a left-handed skew or a right-handed skew depending on which way round the tool is held. Consequently, the numbering system allocates only one number to this tool, although it may be thought of as two.

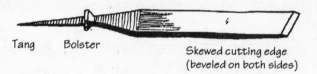

Tang   Bolster

Skewed cutting edge (beveled on both sides)

• **Straight Gouges** (including **Fluters** and **Veiners**) **(3-11)**
Woodcarving gouges are made in as many as nine differ-ent sweeps (the sweep indicating the tightness of the curve); the different Sheffield numbers refer to various sweeps; No. 10 gouges, with a curve that approximates

1	Firmer Chisel
2	Skew Chisel
3-11	Straight Gouges
12-20	Long-bend Gouges (also known as "Curved Gouges")
21-23	Short-bend Chisels (also known as "Front Bent Chisels")
24-32	Short-bend Gouges (also known as "Front-bent Gouges")
33-38	Back-bend Gouges (also known as "Back-bent Gouges")
39, 41, 45	Straight Parting Tools
40, 42, 46	Long-bend Parting Tools (also known as "Curved Parting Tool"')
43-44	Short-bend Parting Tools (also known as "Front-bent Parting Tools")
48	Fishtail Chisel
48G	Fishtail Gouge
50	Box Tool

The Sheffield Numbering System

a U-shape, are known as fluters; No. 11 gouges, with an even deeper shape, are called veiners.

Tang   Outside bevel

• **Long-bend Gouges (Curved Gouges) (12-20)**
Gouges, of various sweeps, with lateral profiles forming a long shallow curve are known as long-bend or curved gouges.

Tang

• **Short-bend Chisels (Front-bent Chisels) (21-23)**
Chisels classified as short-bend or front-bent are also known as "spoon chisels" and "spade chisels." Their shape is designed to make deep cutting easier. No. 21 is straight; No. 22 has a right-hand skew; and No. 23 has a left-hand skew.

Tang   Straight cutting edge

• **Short-bend Gouges (Front-bent Gouges) (24-32)**
Like *short-bend chisels*, short-bend or front-bent gouges may also be referred to as "spoon gouges" and "spade gouges." Nos. 24 to 30 have increasingly tight sweeps. No. 31 is a fluter, and No. 32 is a veiner (*see under Straight Gouges*).

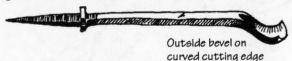

Outside bevel on
curved cutting edge

• **Back-bend Gouges (Back-bent Gouges) (33-38)**
Back-bend or back-bent gouges are particularly useful for fine carving where there is a lot of undercutting to be done, and a front-bend tool would not be able to reach.

Outside bevel

• **Straight Parting Tools (39, 41, 45)**
The parting tool is also called a "V-tool" since it has a V-shaped cross-section. The three numbers refer to increasingly wider V-angles. The parting tool is generally the first tool used when outlining an area to be relief-carved.

Tang   V-shaped cutting edge

• **Long-bend Parting Tools (Curved Parting Tools) (40, 42, 46)**
Like the *straight parting tool*, the long-bend or curved parting tool is made with three different cross-sections—all "V"s of varying angles.

Tang

• **Short-bend Parting Tools (43, 44)**
Short-bend parting tools are also known as "spoon parting tools." These two tool types (Nos. 43 and 44) complete the selection of parting tools and also represent the last carving tools originally defined by the Sheffield List. Subsequent numbers represent continuing modern additions to the system.

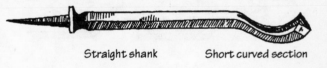

Straight shank   Short curved section

• **Fishtail Chisel (48)**
In general, woodcarving chisels over ¾ in. wide are no longer made with straight sides, but are formed instead toward the cutting edge with rapidly flaring sides, and are referred to as "fishtail" tools.

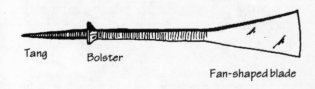

Tang   Bolster
            Fan-shaped blade

• Fishtail Gouge (48G)

Especially large gouges are very often made in the fishtail pattern, and, moreover, are also made with different blade angles known as the "Swiss" style, and the "Tyroler" style.

• Box Tool (50)

The box tool is a broad-bottomed carving tool with upturned sides. It exists with three profiles known respectively as the "macaroni," the "fluteroni," and the "backeroni."

Chisel Gauge

The chisel gauge is a device used chiefly in blind nailing. By attaching the gauge to a ¼–in. *firmer chisel*, with the beveled edge uppermost, a shaving of any thickness can be raised, and when glued down again (a nail having been driven in under the raised shaving), will fit its recess so perfectly that the nail will be invisible or "blind."

Shaving   Chisel

Nail

Chisel Grinder

A chisel grinder is a device for holding a *chisel* or plane iron (the blade of a *plane*) at the correct angle when being rolled back and forth across a *sharpening stone*.

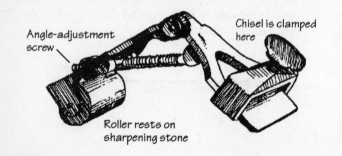

Angle-adjustment screw

Chisel is clamped here

Roller rests on sharpening stone

Chiv: see *Plane, Special-purpose*.

Chute Board (Br. Shooting Board)

A chute board (known in England as a shooting board, since the operation performed with it is known as shooting), was originally no more than two boards fixed together, one wider than the other. A board whose edge needs to be "shot" (planed true and square), is laid in the rabbet formed by the two unequal boards, and a *plane* is worked on its side as illustrated. Assuming the plane itself is perfectly square, the edge of the work will also be planed perfectly square.

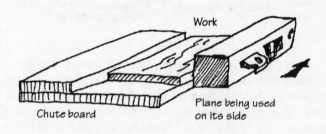

Work

Chute board

Plane being used on its side

Before the advent of the power jointer rendered them obsolete, chute boards had become very sophisticated, made with all manner of fine adjustments and fitted with various holding devices. Metal versions even included a built-in track in which a specially designed chute-plane

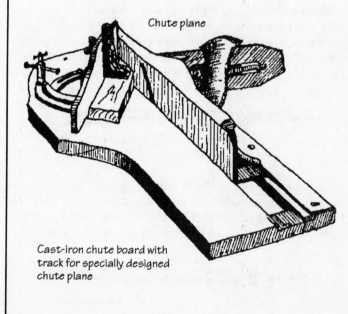

Chute plane

Cast-iron chute board with track for specially designed chute plane

rode, making truing the edge of a board (a procedure requiring considerable skill by hand) an almost fail-proof operation.

Circle Cutter
The circle cutter is a tool that may be used in a *brace* to cut large holes in wood. It is fully adjustable and can cut holes ranging from ³⁄₁₆ in. to 8½ in. in diameter.

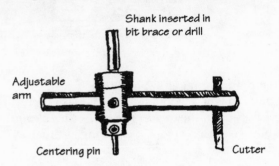

Shank inserted in bit brace or drill

Adjustable arm

Centering pin

Cutter

Circular Plane: see *Plane, Special-purpose.*

Clamp (Br. Cramp)
A clamp is a device used to hold pieces of wood closely and tightly together while being glued or otherwise worked on.

• Clamp, Bar
There are three main types of bar clamp: wooden bar clamps, also known as "sash clamps"' metal bar clamps, and pipe clamps. The wooden variety are no longer common since they cannot compare in strength to metal clamps, although a hybrid variety is sometimes formed by using removable metal parts on a wooden beam or bar.

The bar clamp illustrated is a factory-made transitional model using metal head and foot parts on a wooden beam.

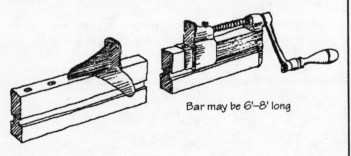

Bar may be 6'–8' long

• Clamp, C- (Br. G-cramp)
This clamp gets its name from its shape, the British name: "G-cramp," being perhaps a little more accurate. These clamps exist in an enormous range of sizes, from tiny 1–in. models to giant 3–ft. examples.

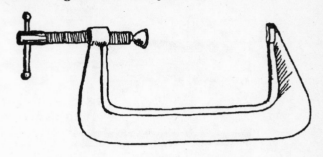

• Clamp, Corner
The corner clamp is actually a form of miter clamp, useful for holding two pieces to be mitered together at right angles—such as picture frames.

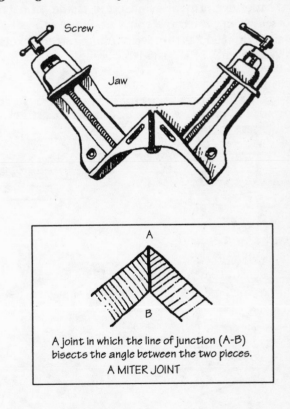

Screw

Jaw

A

B

A joint in which the line of junction (A-B) bisects the angle between the two pieces.
A MITER JOINT

• Clamp, Handscrew
A handscrew is a medium-size clamp consisting of two

wooden jaws joined by parallel screws (called spindles), which rotate in opposite directions.

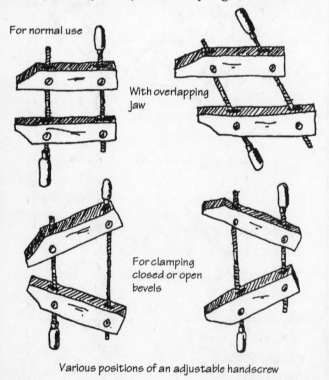

A superior form of handscrew is the adjustable handscrew, in which the spindles, made of steel and threaded right and left are set in rotating steel sockets so that the jaws may be adjusted to any angle, as illustrated.

Various positions of an adjustable handscrew

• **Clamp, Miter-planing** (Br. **Mitre Shooting Block**)
By clamping a workpiece cut to form a miter joint between the jaws of this clamp, and using those jaws as a base for a *plane*, the ends of the wood thus clamped may be shot (planed clean and true) so as to ensure a perfect fit.

• **Clamp, Pianomaker's Wooden**
The pianomaker's wooden clamp is essentially a much larger version of the *cabinetmaker's wooden clamp*, and is capable of clamping work measuring as much as 8 ft. in any direction.

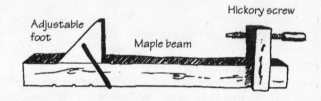

• **Clamp, Sash**
This was the name by which *bar clamps* were known before the introduction of metal bar clamps. The beam is usually square and made of maple, the head is fitted with a hickory screw. Lengths vary from 2 ft. to 6 ft.

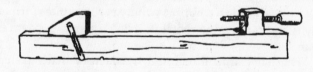

• **Clamp, Spring**
Spring clamps are small, lightweight metal clamps handy for a variety of small jobs. They are made in sizes ranging from 4 1/16 in. to 8 1/2 in.

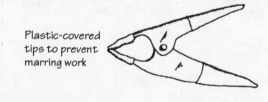

Plastic-covered tips to prevent marring work

• **Clamp, Web**
Rope tied around irregularly shaped objects and then tightened in the manner of a tourniquet, has long been used as a form of clamp. With the advent of nylon, the modern web clamp was developed. It consists of a 12–ft. nylon web and a metal fastener.

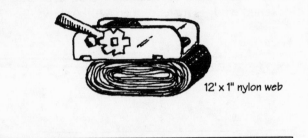

12' x 1" nylon web

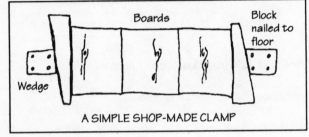

Boards
Block nailed to floor
Wedge
A SIMPLE SHOP-MADE CLAMP

Clapboard Gauge: see *Gauge.*

Cluster-bead Beading Plane: see *Plane, Moulding.*

Cold Chisel: see *Chisel, Cold.*

Compass
The compass is sometimes referred to in the plural as a "pair of compasses." The singular is to be preferred, however, as being simpler and more accurate. A compass in its simplest form is merely a pair of equal-length legs, usually metal, joined at one end. Its function is to describe circles, and to this end it may be made entirely of metal, the points of the legs being sharpened so as to form a scribe. Alternatively, one of the legs may have

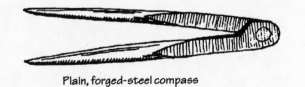

Plain, forged-steel compass

provision for holding a separate marking instrument, such as a pen or a pencil.

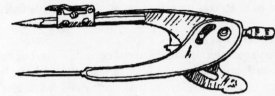

Lightweight pencil compass

Compass Plane: see *Plane, Special-purpose.*

Compass Saw: see *Saw.*

Cooper's Adze: see *Adze.*

Cooper's Chiv Plane: see *Plane, Special-purpose.*

Cooper's Croze Plane: see *Plane, Special-purpose.*

Cooper's Driver: see *Hammer.*

Cooper's Long Jointer: see *Plane, Special-purpose.*

Cooper's Sun Plane: see *Plane, Special-purpose.*

Coping Saw: see *Saw.*

Core Box Plane: see *Plane, Special-purpose.*

Corner Chisel: see *Chisel, Wood, Bench.*

Cornering Tool
The cornering tool is 5 ½ in. long and is made in two sizes, each having two different size holes: ¹⁄₁₆ in. and ⅛ in., and ¼ in. and ⅜ in. It is used for taking the edge off sharp corners and giving them a small radius.

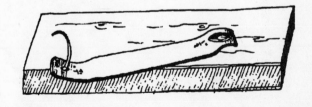

Countersink

A countersink is a tool that makes a conical enlargement of the upper part of a hole for receiving the head of a screw or bolt. It is usually operated in a *drill* or *brace*.

• Countersink, Rose

The rose countersink is the most common kind of countersink (designed for use with handtools), although the design is very similar to that employed with electric drills.

• Countersink, Snail

This variety has only one deep, spiral cutting edge, which gives it somewhat the appearance of a snail's shell.

Cove Plane: see *Plane, Moulding.*

Cramp: see *Clamp.*

Crosscut Saw: see *Saw.*

Crow Bar

The crow bar, originally called simply a "crow," is a large bar of iron. Old crows were used as levers for lifting heavy beams, and one end was furnished with a claw. Modern crow bars are made of cast steel, and have thickened wedges instead of crow-like claws.

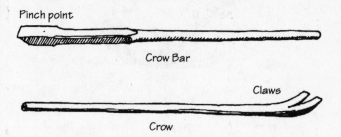

Pinch point

Crow Bar

Claws

Crow

Crown Moulding Plane: see *Plane, Moulding.*

Croze: see *Plane, Special-purpose.*

Curved Gouge: see *Chisel, Wood, Woodcarving.*

Curved Parting Tool: see *Chisel, Wood, Woodcarving.*

Cutting Gauge: see *Gauge.*

Saw Grinding, from *The Illustrated Sheffield Guide,* 1879

The Illustrated Encyclopedia of Woodworking Handtools Instruments & Devices

The Cooper, from *Das Ständebuch* (The Book of Trades), by Jost Amman, 1568

Tools shown include: cooper's jointer, dividers, bow saw, mallet, and cooper's dog.

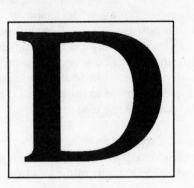

Dado Plane: see *Plane, Special-purpose.*

Deck Chisel: see *Chisel, Wood, Special-purpose.*

Diamond-nose Chisel: see *Chisel, Cold.*

Dividers
Dividers developed from the *compass*, and are called properly—though rarely—"dividing compasses." They are, in fact, a pair of compasses (more usually and simply referred to as a "compass") with steel points at the end of each leg. Used for dividing lines, transferring dimensions, and general measuring, they are often made with the same methods of adjustment as *calipers*.

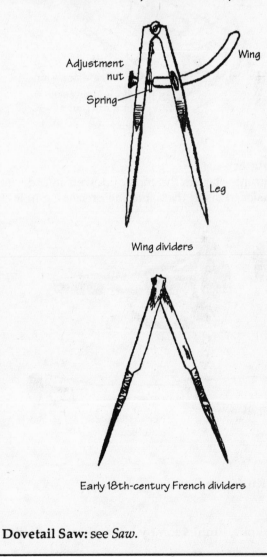

Wing dividers

Early 18th-century French dividers

Dovetail Saw: see *Saw.*

Doweling Jig

A doweling jig is the name given to any one of a number of differently designed, patented devices for accurately controlling the boring of holes for dowels when making blind dowel joints (*see box on Doweling Jig Uses*).

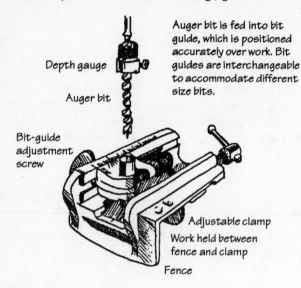

Depth gauge

Auger bit is fed into bit guide, which is positioned accurately over work. Bit guides are interchangeable to accommodate different size bits.

Auger bit

Bit-guide adjustment screw

Adjustable clamp

Work held between fence and clamp

Fence

Dowel Pointer

A dowel pointer tapers the ends of dowels in order to make it easier to insert them into their receiving holes.

Adjustable dowel pointer for use in a bit brace

Wooden-handled dowel pointer

Drawing Knife: see *Knife*.

Drawknife

The drawknife, until recently, was always called a "drawing knife" because it is simply a knife, with a handle at each end, operated by being drawn towards the user. Also known as a "draft shave" and a "draw shave" (which points to its relationship with the *spokeshave*), it was an extremely popular tool because of its ability to remove a lot of wood very quickly. Consequently, many different kinds of drawknives have been developed, each adapted to the particular requirements of various woodworking trades.

• Drawknife, Carpenter's

This is the standard drawknife, still in use today when a freehand two-handled trimming tool is needed.

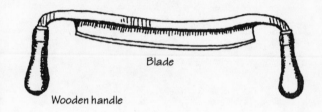

Blade

Wooden handle

• Drawknife, Carriage-body

The carriage-body drawknife, while it adheres to all the requirements necessary for being called a drawknife, is the least like all other drawknives because its blade, which is interchangeable, is only 1 ½ in. wide.

Note nut securing interchangable blade

Blades may be ½", 1", or 1½" wide.

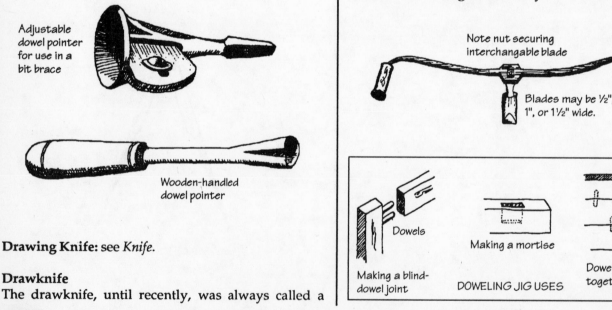

Dowels

Making a blind-dowel joint

Making a mortise

DOWELING JIG USES

Doweling boards together

• Drawknife, Carriagemaker's

The carriagemaker's drawknife has a much straighter profile than the *carpenter's drawknife*, as well as having a different blade cross-section (*see box on Drawknife Cross-sections*).

• Drawknife, Chamfer Knife

Originally intended for use when forming large chamfers on heavy beams, the chamfer knife is distinguished by having one of its handles formed at right angles to the other.

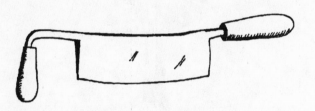

• Drawknife, Coachmaker's

This drawknife, with a relatively short and curved cutting edge, was preferred by coachmakers.

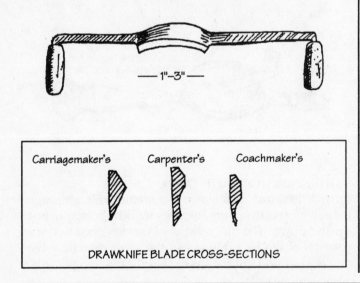

— 1"–3" —

• Drawknife, Cooper's

The cooper's drawknife is somewhat narrower than the *carpenter's drawknife*, consisting of a straight blade with splayed handles.

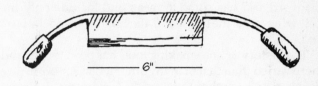

— 6" —

• Drawknife, Jigger

The jigger is another variety of *coachmaker's drawknife*, and has a combination blade that can make both straight and hollow cuts. Coopers also used a drawknife called a jigger, made with the curved section of blade only, which was used when a suitable *chiv plane* was unavailable.

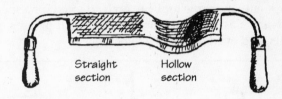

Straight section   Hollow section

• Drawknife, Mast

The mast drawknife is the biggest of all drawknives, often having a blade as long as 24 in.

• Drawknife, Wagonmaker's (Br. Waggonwright's Drawknife)

Wagonmakers used the heaviest of the regular drawknives, the blades being 1 ¼ in. wide and 10 in. broad.

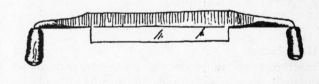

DRAWKNIFE BLADE CROSS-SECTIONS

Carriagemaker's   Carpenter's   Coachmaker's

Drill (1)

In woodworking a drill is an instrument for boring a hole, and so the word *drill* might apply equally to all the different tools that exist to do this, such as the *brace*, the *auger*, the *gimlet*, and many more. Nowadays, however, the word *drill* has come to mean and include only the following tools: the *breast drill*, the *hand drill*, and the *push drill*.

The history of hole-making tools has been longer and more varied than that of many other tools. For centuries, simple devices such as the strap drill, the pump drill, and the bow drill were the only tools for making holes. Although the *brace* was known by the Middle Ages, it was not until 1770 that the principal of the spiral auger was realized by Phineas Cooke in England. Hand drills and breast drills were not invented until the middle of the nineteenth century. Nevertheless, within fifty years there were a great many varieties of the basic hand drill—all operating under the same principle and all essentially the same tool, but each adapted for a specific use—such as the bench drills, bench drill presses, post drills, and even boring machines.

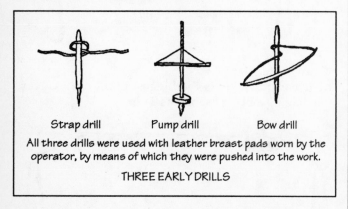

Strap drill   Pump drill   Bow drill

All three drills were used with leather breast pads worn by the operator, by means of which they were pushed into the work.

THREE EARLY DRILLS

• Drill, Breast

The breast drill is constructed on the same principle as the *hand drill*, but is intended for drilling much bigger holes. Since this requires more pressure than can be exerted on a small drill, the breast drill is designed to be held in a pad against the chest when being used, allowing more force to be applied.

• Drill, Hand

While *breast drills* will accept *bits* up to ½ in. in diameter, hand drills are generally made for use only with smaller

bits with shanks of ¼ in. or ⅛ in. diameter. Better quality hand drills are fitted with gearing mechanisms for producing different rotational speeds.

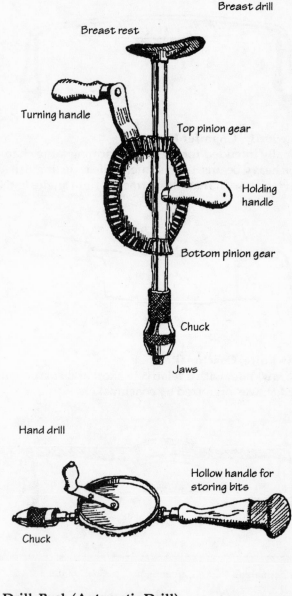

Breast drill

Breast rest

Turning handle

Top pinion gear

Holding handle

Bottom pinion gear

Chuck

Jaws

Hand drill

Hollow handle for storing bits

Chuck

• Drill, Push (Automatic Drill)

Push drills are also known as automatic drills, although they are not really automatic; they work by being pushed up and down. The older variety is so designed that the rotation of the *bit* is always in the same direction; the newer kind's bit reverses as the handle moves upwards.

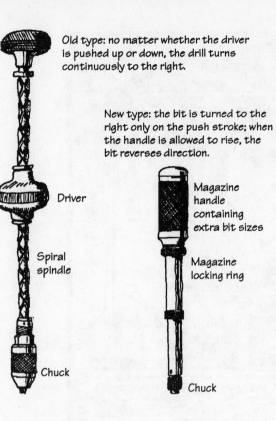

Old type: no matter whether the driver is pushed up or down, the drill turns continuously to the right.

New type: the bit is turned to the right only on the push stroke; when the handle is allowed to rise, the bit reverses direction.

Driver

Spiral spindle

Chuck

Magazine handle containing extra bit sizes

Magazine locking ring

Chuck

Drill (2)
'Drill' is the word used by the machinist and metal-worker to designate the tool properly called by the woodworker a *bit*. In fact, the term "machinist's drill" usually implies a *twist bit*.

Drill Stop
The drill stop is to the *twist bit* what the *auger bit gauge* is to the *auger bit*; that is, a device for controlling the depth to which the *twist bit* may drill.

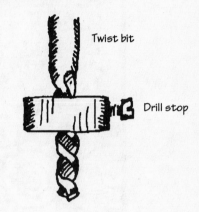

Twist bit

Drill stop

The Turner, from *Das Ständebuch* (The Book of Trades), by Jost Amman, 1568

Emery Wheel

An emery wheel is a wheel-shaped mass of hard, granular material called emery, used for grinding and sharpening metal tools. It is typically mounted in a frame so that it may be rotated, either by hand or electricity.

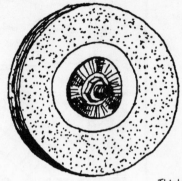

Thickness: ¼"–4"
Diameter: 1"–36"

Emery Wheel Dresser

This tool is used for shaping the *emery wheel* when, in the course of use, it has become grooved or worn out of round. Its cutters are capable of resisting great wear since they are specially hardened during their manufacture.

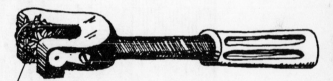

Cutters

Engineer's Hammer: see *Hammer*.

Expansive Bit: see *Bit*.

Extension Bit Holder: see *Bit Holder*.

The Lutemaker, from *Das Ständebuch* (The Book of Trades), by Jost Amman, 1568

Tools shown include: jack plane, smooth plane, glue pot, chisel, mallet, scrapers, and axe.

Farrier's Hammer: see *Hammer*.

Felt Knife: see *Knife*.

Fiber-head Mallet: see *Mallet*.

File

A file is a hardened steel instrument, its surfaces covered with parallel rows of sharp teeth or furrows, generally at an angle of 65° to the center line. Files are used for shaping and smoothing wood, metal, and other materials. The following discussion focuses mainly on those files used by the woodworker.

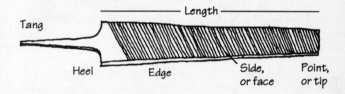

Files are classified by the type and coarseness of their cut, their cross-section and longitudinal shape, and their length. Their width usually increases proportionately as they become longer.

The Cut: Files may be single-cut, double-cut, or curved-cut, as illustrated. Single-cut files are used for producing fine, smooth finishes. Double-cut files are used for faster removal of material, and consequently produce a rougher cut than single-cut files. Curved-cut files have a much coarser cut, and are used mainly for lead or the alloy, babbitt. The cuts are also graded so that each type may have more or fewer teeth. This is called the degree of coarseness. The main grades, or degrees of coarseness,

Single-cut   Double-cut   Curved-cut

are: coarse, bastard, second cut, and smooth. Each of these degrees of coarseness may be single-cut or double-cut.

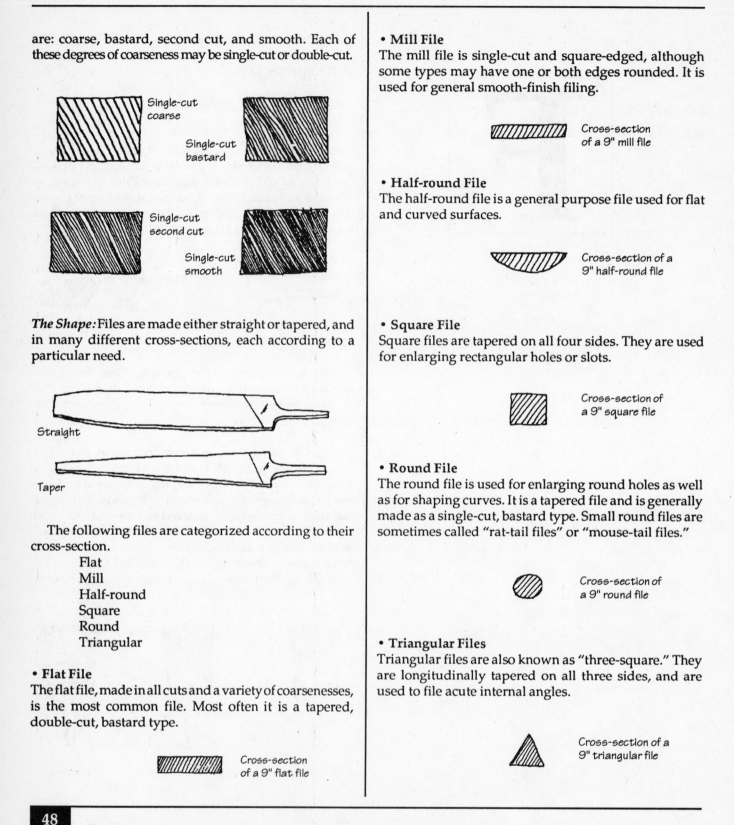

Single-cut coarse

Single-cut bastard

Single-cut second cut

Single-cut smooth

The Shape: Files are made either straight or tapered, and in many different cross-sections, each according to a particular need.

Straight

Taper

The following files are categorized according to their cross-section.

Flat
Mill
Half-round
Square
Round
Triangular

• Flat File
The flat file, made in all cuts and a variety of coarsenesses, is the most common file. Most often it is a tapered, double-cut, bastard type.

Cross-section of a 9" flat file

• Mill File
The mill file is single-cut and square-edged, although some types may have one or both edges rounded. It is used for general smooth-finish filing.

Cross-section of a 9" mill file

• Half-round File
The half-round file is a general purpose file used for flat and curved surfaces.

Cross-section of a 9" half-round file

• Square File
Square files are tapered on all four sides. They are used for enlarging rectangular holes or slots.

Cross-section of a 9" square file

• Round File
The round file is used for enlarging round holes as well as for shaping curves. It is a tapered file and is generally made as a single-cut, bastard type. Small round files are sometimes called "rat-tail files" or "mouse-tail files."

Cross-section of a 9" round file

• Triangular Files
Triangular files are also known as "three-square." They are longitudinally tapered on all three sides, and are used to file acute internal angles.

Cross-section of a 9" triangular file

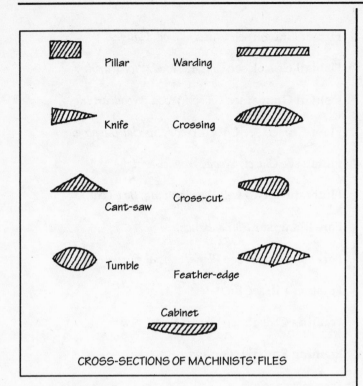

CROSS-SECTIONS OF MACHINISTS' FILES

Pillar   Warding   Knife   Crossing   Cant-saw   Cross-cut   Tumble   Feather-edge   Cabinet

• File, Needle

Needle files are files smaller even than *die-sinkers' files.* Both types are sometimes referred to as "Swiss-pattern files." Needle files are used mainly by jewelers or watch-makers, but are sometimes useful for woodworkers.

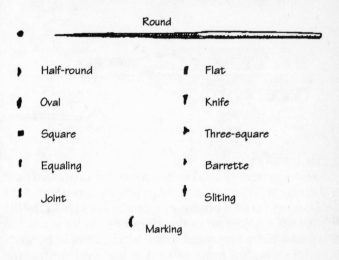

Round

Half-round   Flat

Oval   Knife

Square   Three-square

Equaling   Barrette

Joint   Sliting

Marking

• File, Auger-bit

The auger-bit file is a small file designed especially for filing and sharpening *auger bits.* It has alternate safe (uncut) edges, allowing the different surfaces of the file to be used without fear of the adjacent edges inadvertently doing any damage to areas not needing to be filed.

• File, Woodcarver's

The woodcarver's file is a *riffler* with a surface cut more like a file. A riffler is a curved tool used by carvers and sculptors, usually finished like a *rasp* (a tool similar to a file but having a much coarser cutting surface). The woodcarver's file is therefore also known as a "riffler file." Like other files, woodcarver's files are made in a variety of shapes and cross-sections.

• File, Die-sinker's

A die-sinker's file is a small file, about 3½ in. long, used to file small instruments and machine parts. The files whose cross-sections are illustrated here are all tapered to a sharp point.

Flat (1 safe edge)   Lozenge   Half-round   Oval

Flat   Three-square   Knife   Oval (1 sharp edge)

Square   Round   Auriform   Oval (2 round edges)

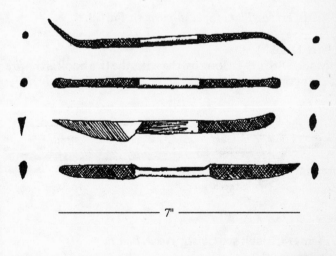

7"

File Brush

During filing, the teeth of a *file* may clog up with particles of the material being worked on. To help prevent this, chalk is often rubbed between the teeth prior to filing, but ultimately the file must be cleaned. This is the job of the specially designed file brush.

File carding (fine wire bristles)

Slot for pick

Brush

Pick

File Holder

Files are generally used with wooden handles fitted over their tangs. When a file is being used for surface smoothing, however, it is often more convenient to hold the file with a file holder since this allows the entire surface of the tool to be presented to the work. The file holder's handle is hollow and threaded so that the length of the rod may be adjusted to fit different length files.

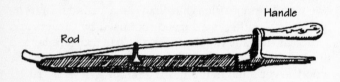

Handle

Rod

Filletster Plane: see *Plane, Special-purpose, Rabbet.*

Fillister: see *Plane, Special-purpose, Rabbet.*

Fire Hook

Made with a 6–ft. long ash handle, the fire hook is the *fire axe's* companion.

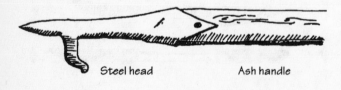

Steel head

Ash handle

Firmer Chisel: see *Chisel, Wood, Bench.*

Firmer Gouge: see *Chisel, Wood, Gouges.*

Fishtail Chisel: see *Chisel, Wood, Woodcarving.*

Fishtail Gouge: see *Chisel, Wood, Woodcarving.*

Floor Chisel: see *Chisel, Wood, Special-purpose.*

Fluter: see *Chisel, Wood, Woodcarving.*

Fluteroni: see *Chisel, Woodcarving, Box Tool.*

Fore Plane: see *Plane, Bench.*

Forkstaff Plane: see *Plane, Special-purpose.*

Forstner Bit: see *Bit.*

Framing Chisel: see *Chisel, Wood, Bench.*

Framing Square: see *Square.*

Fretsaw: see *Saw.*

Froe

The word *froe,* or, as it used to be spelled, *frow,* comes from the now archaic word: *froward*—which was used as the opposite of *toward,* meaning "turning away from." This is exactly what the froe does; it is a wedge-shaped tool, which, when driven into the wood and moved back and forth, "turns the wood away" from the worker, and so splits off a section. Since froes split wood along the grain easily, they were used for many jobs, from splitting (or "riving") blocks of cedar, pine, or redwood into roofing shingles, to making lath for plastering.

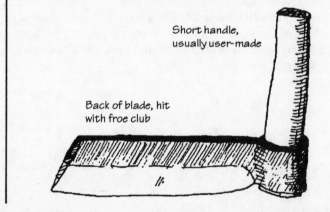

Short handle, usually user-made

Back of blade, hit with froe club

• **Froe, Cooper's**
A common adaptation of the basic froe was the cooper's curved froe, which was used for making curved barrel staves.

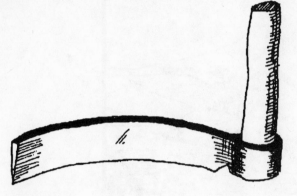

• **Froe, Knife**
Froes are normally characterized by having the handle

set at right-angles to the blade—an arrangement that provides a little more leverage to the handle when the tool is being twisted. The knife froe is an unusual example of a froe, with its handle in the same plane as the blade. It, too, is hit with a *mallet* on the back of the blade.

Froe Club: see *Mallet*.

Front-bent Chisel: see *Chisel, Wood, Woodcarving*.

Front-bent Gouge: see *Chisel, Wood, Woodcarving*.

Front-bent Parting Tool: see *Chisel, Wood, Woodcarving*.

Carpenters Framing a Building, from *Perspectiva*, by Hieronimus Rodler, 1546

Tools shown include: trimming axe (being used), broad axe, saw horse, pinch dog, sword saw, spoon auger, chisel, mallet, square, compass, chalk line, and two-man crosscut saw.

Gauge

The origin of the word *gauge*, sometimes spelled *gage*, is unknown. It has many meanings, but, as far as woodworking is concerned, it refers to various instruments used for measuring, marking, or testing.

• Gauge, Butt

A butt gauge is a gauge used to mark the position of butts (hinges) to be installed when hanging doors and windows.

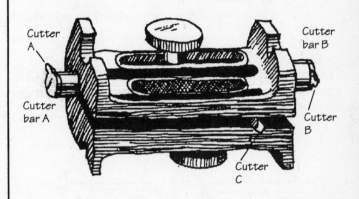

Cutter bar *A* has two cutters mounted on it. When cutter *A* is set for gauging on the edge of the door, cutter *C*, being on the same bar, is automatically set for gauging from the back of the jamb. Cutter bar *B* has a steel cutter, cutter *B*, to gauge accurately the thickness of the butt. The butt gauge is thus a *rabbet gauge*, a *marking gauge*, and a *mortise gauge* all in one, and is therefore capable of being used for all door trim, including lock plates, strike plates, and other hardware.

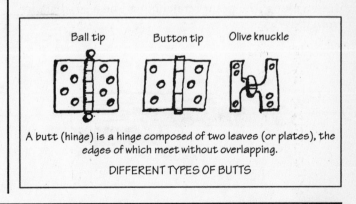

A butt (hinge) is a hinge composed of two leaves (or plates), the edges of which meet without overlapping.

DIFFERENT TYPES OF BUTTS

• Gauge, Clapboard

The clapboard gauge is used in applying clapboarding to ensure that each board overlaps the previous board by the proper amount.

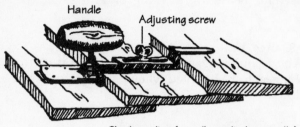

Clapboarding (usually applied vertically)

• Gauge, Cutting

The cutting gauge is a cross between a *splitting gauge* and a *marking gauge*. It is fitted with a small knife-like blade that cuts a clean line across the surface of the work.

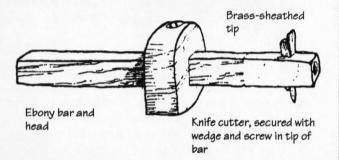

• Gauge, Marking

The marking gauge is used for marking a line parallel with the edge of a piece of wood. Early marking gauges were made from one piece of wood and a small nail. The shoulder was held against the edge of the work and drawn along, the point scratching a line a set distance from the edge.

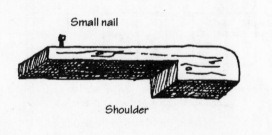

The one-piece marking gauge could only be used for a single measurement, and consequently it disappeared with the introduction of the adjustable gauge, which consists of a movable head (or stock) on a bar (or beam).

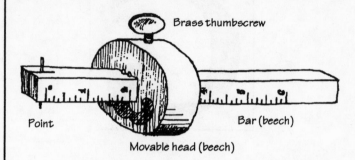

The marking gauge illustrated above is used mainly for wood. The metal marking gauge shown below may be used for metal or wood.

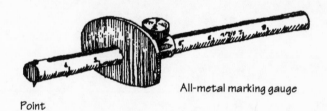

• Gauge, Mortise

The mortise gauge is constructed similarly to the *marking gauge*, but with two points instead of one. The second point is adjustable so that two lines at varying distances from one another may be scribed at varying distances from the edge of the work. This is necessary when laying out mortises and tenons of different sizes (*see box on page 55*).

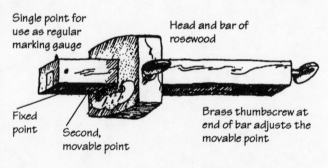

Like the *marking gauge*, the mortise gauge is made in a more durable metal model. The metal mortise gauge, however, is not as convenient as the wooden model because two operations are necessary to scribe the double line, since the points are on separate bars.

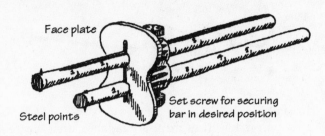

Face plate
Steel points
Set screw for securing bar in desired position

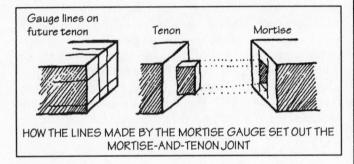

Gauge lines on future tenon
Tenon
Mortise

HOW THE LINES MADE BY THE MORTISE GAUGE SET OUT THE MORTISE-AND-TENON JOINT

• Gauge, Panel

The panel gauge is a large *marking gauge* used for much bigger work.

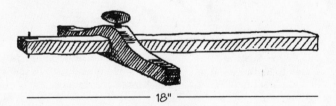

18"

• Gauge, Rabbet (Br. Rebate Gage)

The rabbet gauge may also be used as a *butt gauge*, and as such is useful when hanging doors, mortising, and marking all kinds of joints and grooves in wood.

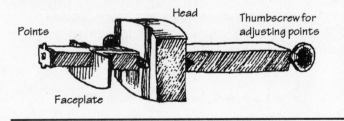

Points
Head
Thumbscrew for adjusting points
Faceplate

• Gauge, Scratch

The scratch gauge is a very exact *marking gauge*, graduated in ⅟₆₄ in., for use on metal as well as wood. Its cutters are replaceable.

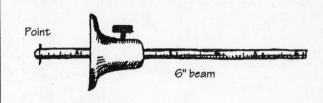

Point
6" beam

• Gauge, Slitting

A slitting gauge is a device for accurately slitting thin boards. The adjustable head regulates the width of the wood being split off, and a roller set in the base of the handle (which is necessarily pressed against the work) allows the gauge to move along smoothly.

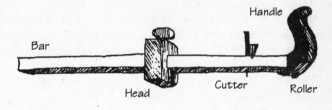

Bar
Handle
Head
Cutter
Roller

G-clamp: see *Clamp*.

Gent's Saw: see *Saw*.

Gimlet

The gimlet, sometimes spelled "gimblet" and also known as a "wimble," is a miniature *auger*. Its purpose is to bore small holes in wood. There are two kinds: spiral-grooved and straight-grooved, both made in a range of diameters from ⁴⁄₃₂ in. to ⁸⁄₃₂ in.

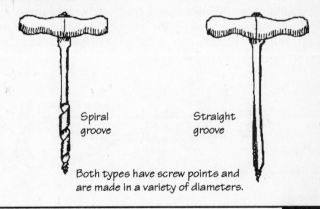

Spiral groove
Straight groove

Both types have screw points and are made in a variety of diameters.

Glass Cutter

A glass cutter is a device for scoring glass that may then be broken cleanly along the score. Inasmuch as many pieces of woodwork are glazed, such tools are often used by woodworkers, although they are properly glaziers' tools.

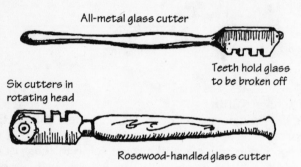

All-metal glass cutter

Teeth hold glass to be broken off

Six cutters in rotating head

Rosewood-handled glass cutter

• Glass Cutter, Circular

The regular glass cutter is operated by being drawn along a guiding straightedge. Consequently it can cut only in straight lines. For cutting curved sections the circular glass cutter is used.

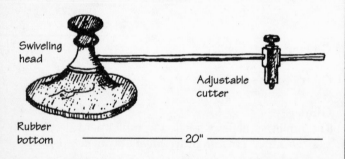

Swiveling head

Adjustable cutter

Rubber bottom

20"

• Glass Cutter, Gauge

The gauge glass cutter needs no *straightedge* to help it score a straight line, since the gauge may be adjusted to any width and held against the edge of the glass in the same manner as a *marking gauge*.

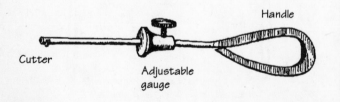

Handle

Cutter

Adjustable gauge

• Glass Cutter, Glazier's Diamond

A diamond is very hard, and is used for the best quality glass cutters instead of the regular hardened-metal cutter wheel. This kind of glass cutter is often referred to simply as a "glazier's diamond."

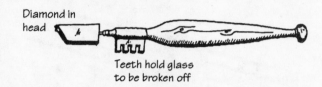

Diamond in head

Teeth hold glass to be broken off

Glue Pot

Many man-made glues are now available, such as casein glue, epoxy glue, polyvinyl resin glue, and resorcinol glue. These have largely replaced the traditional animal-hide glue except in high-quality repair and veneer work. (The advantage of the traditional glue is that it may be easily undone should subsequent repair be necessary.) Although electrically heated, thermostatically controlled glue pots are now common, the type of glue pot illustrated here, in which the sheets, flakes, or powdered material are melted and prepared for use, is still preferred by many workers.

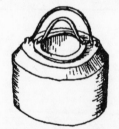 

Iron glue pot, tinned inside

Different model, showing how inner pot is warmed by water heated in outer pot

Gouge: see *Chisel, Wood, Gouges*.

Grindstone

A modern grindstone no longer consists of natural stone (such as sandstone) but is usually made from electrically baked abrasive materials, and is typically driven by electricity. Older grindstones were hand or foot driven, their purpose being to grind and shape metal edges prior to their being sharpened on a *sharpening stone*.

Gunsmith's Screwdriver: see *Screwdriver*.

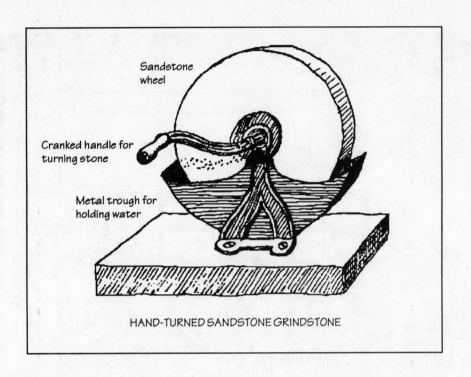

Sandstone
wheel

Cranked handle for
turning stone

Metal trough for
holding water

HAND-TURNED SANDSTONE GRINDSTONE

Medieval Italian Carpenters

Tools shown include: framed pit saw, chalk line, jointer plane, mallet, and adze (all in use);
chisel, square, smooth plane, and hammer (in basket).

Hacking Knife: see *Knife*.

Hacksaw: see *Saw*.

Hacksaw: see *Saw*.

Half-roundnose Chisel: see *Chisel, Cold*.

Hammer

A hammer is one of the most basic and oldest tools of mankind. It was originally probably just a heavy stone, as suggested by its etymology, which relates the common Teutonic root: *hamor* to the Slavic word: *kamy*, meaning "stone." A hammer essentially consists of a heavy head, usually of metal, set crosswise on a handle, and is used in woodworking primarily for driving nails as well as for other beating and breaking operations.

• Hammer, Adze-eye Bell-face

The adze-eye bell-face hammer is an example of the extreme specialization that often develops with a much-used tool. It is basically a *nail hammer*, used by carpenters for driving nails. Its long name refers to its special characteristics; "adze-eye" means the haft (handle) is fitted into a sleeve at the base of the head, in the same manner as an *adze*; "bell-face" refers to the fact that the head is slightly rounded so that a nail may be driven below the surface without the wood being dented.

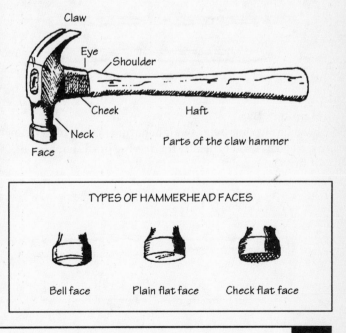

Parts of the claw hammer

TYPES OF HAMMERHEAD FACES

Bell face   Plain flat face   Check flat face

• Hammer, Adze-eye Claw

Similar to the *adze-eye bell-face hammer*, but without the rounded face, the adze-eye claw hammer is intended for rough work where inadvertent denting of the wood is unimportant. Its flat face may be smooth, as on bell-faced hammers, or checked to help prevent the face slipping off the nail being driven.

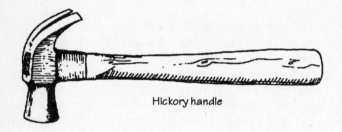

Hickory handle

• Hammer, Bill-poster's Tack

A bill-poster's tack hammer, like a regular *tack hammer*, has a head, one end of which is magnetic. This allows a nail that is too small to be held by the fingers to be driven where desired. The head is usually straight; the handle varies in length from 11 in. to one made in three sections, measuring 45 in. when completely assembled.

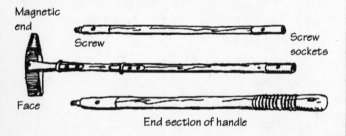

Magnetic end

Screw

Screw sockets

Face

End section of handle

• Hammer, Brad

A brad hammer is a lightweight hammer weighing from 2 oz. to 4 oz., used typically for driving brads (small nails without flat heads).

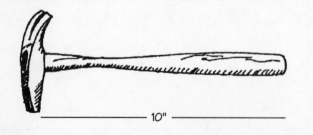

10"

• Hammer, Brick

The brick hammer, also known as a "bricklayer's hammer" is designed for cutting and breaking bricks and concrete blocks. The sharp end used for cutting and scoring is called the bit. These tools are made with both hickory handles and chrome-plated steel-tube handles covered with a contoured plastic or rubber grip.

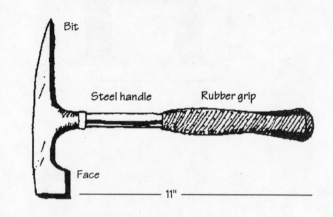

Bit

Steel handle   Rubber grip

Face

11"

• Hammer, Carpet-layer's

The carpet-layer's hammer is a lightweight hammer related to the carpenter's *tack hammer*.

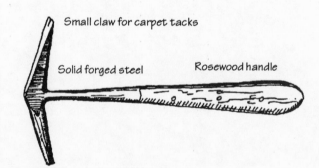

Small claw for carpet tacks

Solid forged steel   Rosewood handle

• Hammer, Cooper's Driver

The cooper's driver is an unusual type of hammer; its handle is a continuation of its head. It is driven like a *socket firmer chisel* rather than being swung in the normal manner of a hammer.

Face   Ring

• Hammer, Copper

A copper hammer is so called because it is made of copper. Used in metalwork, its soft copper head will not mar the work.

1 lb.–2 lb.

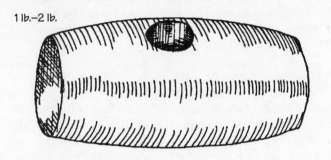

• Hammer, Double-face

The double-face hammer is a small *sledge hammer* used by engineers. It ranges in weight from 24 oz. to 58 oz., its handle averaging 16 in. in length.

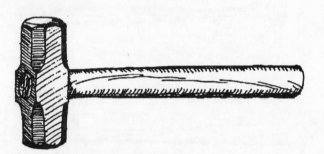

• Hammer, Engineer's

The term *engineer's hammer* may refer to the *double-face hammer* above, but properly means a small cross-pein *sledge hammer*.

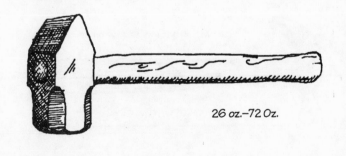

26 oz.–72 oz.

• Hammer, Farrier's

A farrier is one who shoes horses. The shoes are made of iron, hence the name *farrie'* which comes from the Latin word for iron: *ferrum*.

Claw

7 oz. or 13 oz.

Face

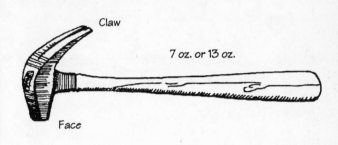

• Hammer, Framing

A framing hammer is a larger cousin to the *nail hammer*. It is used by house framers, carpenters, and construction workers. It weighs at least 20 oz. and sometimes as much as 32 oz. The handle is a little longer and the claw is usually straighter than that of the *nail hammer* since it is often used for ripping.

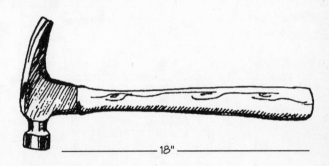

18"

• Hammer, Jeweler's

The jeweler's hammer is a very lightweight *machinist's hammer*. As well as the kind illustrated, jeweler's hammers are also made with a cross pein, similar to the *engineer's hammer*.

Ball pein

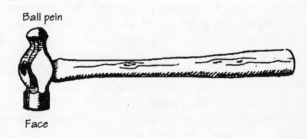

Face

• Hammer, Joiner's

The joiner's hammer is a cross between a *framing hammer* and an *engineer's hammer*. It may be made with either a plain face or a checked face.

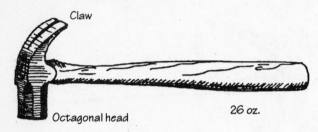

Claw

Octagonal head

26 oz.

• Hammer, Machinist's

The machinist's hammer, also known as a "pein hammer," is used mainly in metalwork.

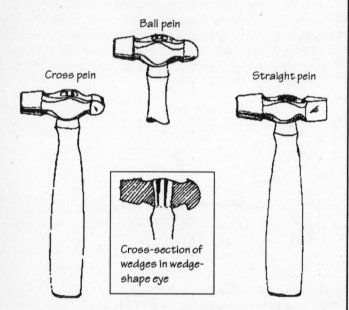

Ball pein

Cross pein

Straight pein

Cross-section of wedges in wedge-shape eye

Pein, Pane, or Peen: When the opposite end from the hammer's face ends in a relatively thin wedge-like shape it is called in Britain a "paned hammer," and in America a "peened hammer." In both places it is now most commonly referred to as a "peined hammer"—pronounced either as *pane* or *peen* depending on the user's preference. The origin of this curious word is obscure, but while there are sound etymological reasons for both *pane* and *peen, pein* (which often causes first-time users pronunciation pains) it is a nineteenth-century manufacturer's invention that might be usefully ignored.

Both cross and straight peins are useful for starting small nails, though their original purpose is for hammering out metal in ways not possible with a round-faced tool. An extension of this function has led to the provision of some hammers with an almost spherical face (at the opposite end from the regular face), known perversely as a "ball-pein"—although "ball-face" would be more accurate since pein is, by definition, wedge-shaped.

• Hammer, Magnetic Tack

The magnetic tack hammer is a lightweight (5 oz.) hammer, the end of the head opposite its face being magnetized in order to hold small tacks that are too tiny to be held easily between the fingers. With the tack held by the magnetized end, the hammer is wielded lightly, fixing the tack upright so that it may then be driven in the normal fashion.

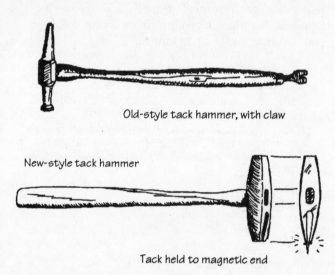

Old-style tack hammer, with claw

New-style tack hammer

Tack held to magnetic end

• Hammer, Nail

The nail hammer is the most common hammer, having been used by carpenters for centuries. Its design has

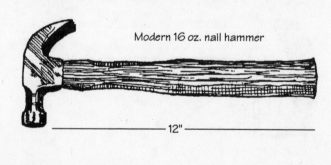

Modern 16 oz. nail hammer

12"

remained virtually unchanged since Roman times. As Peter Nicholson put it in 1812: "The use of the Hammer is for driving nails into wood by percussive force."

Wooden or Metal Handles: Many hammers are made nowadays with steel handles. These are undoubtedly stronger and less liable to break, but for the craftsman they are no substitute for the feel and spring of a well-shaped hickory handle.

• Hammer, Riveting

Riveting hammers are made in a larger variety of sizes than most other hammers, and range from 1½ oz. to 26 oz. They have flat faces and cross peins, and are used for driving, upsetting, and heading rivets (three stages in riveting).

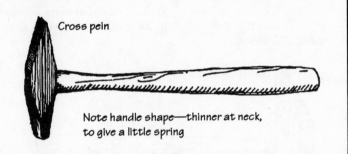

Cross pein

Note handle shape—thinner at neck, to give a little spring

• Hammer, Shoe (Cobbler's Hammer)

Despite its name, the cobbler's hammer—a common eighteenth-century design—was used as much by cabinetmakers as by cobblers. Its oblong end was especially useful for drawer corners.

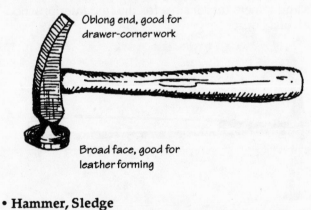

Oblong end, good for drawer-corner work

Broad face, good for leather forming

• Hammer, Sledge

A sledge hammer is a very heavy hammer with a long handle. It may weigh from 2½ lbs. to as much as 24 lbs.; the handle increases in length with the weight from 16 in. to 3 ft.

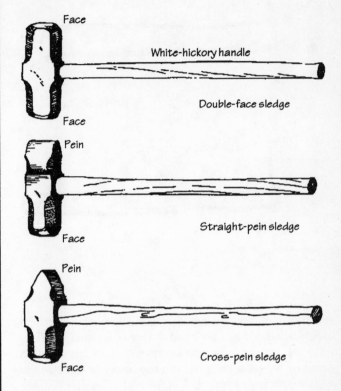

Face

White-hickory handle

Face

Double-face sledge

Pein

Face

Straight-pein sledge

Pein

Cross-pein sledge

Face

• Hammer, Soft-face

Soft-face hammers are used where it is necessary to strike a hard blow without marring or bruising the surface of the work. Today, most soft-face hammers are made of a tough, slow-burning plastic at the actual tip (often replaceable). Older hammers were made of rolled rawhide, which could also be replaced when worn out. They are made in a great variety of weights. Rubber soft-face hammers, much used by other trades (and also for replacing automobile hubcaps) are generally avoided by

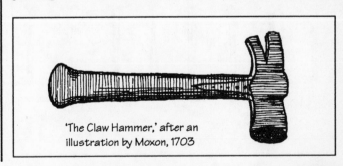

'The Claw Hammer,' after an illustration by Moxon, 1703

the woodworker since the rubber often marks the wood.

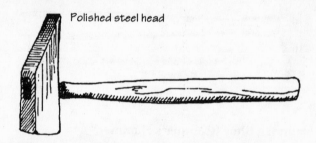

Rawhide head

Plastic head

• Hammer, Striking

A striking hammer is a heavy hammer with a handle as much as 3 ft. long and a head that may weigh from 5 lbs. to 24 lbs. There are two basic types of head: the so-called Long or Nevada Pattern, and the Short or Oregon Pattern (an inch shorter).

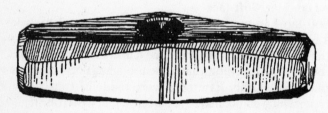

6 ¾" (Long-pattern)

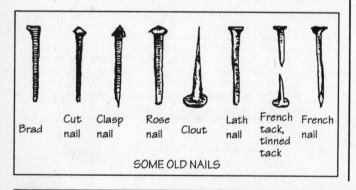

Brad — Cut nail — Clasp nail — Rose nail — Clout — Lath nail — French tack, tinned tack — French nail

SOME OLD NAILS

• Hammer, Tack

Most tack hammers are magnetic (*see Magnetic Tack Hammer*). There are, however, a few that are not.

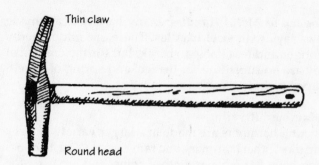

Thin claw

Round head

• Hammer, Tinner's (Tinsmith's Hammer)

This is also called a "paneing (or 'peining') hammer" and a "tinner's setting hammer," since it is used in metalwork for both setting and peining, as well as dressing the seams of tinwork.

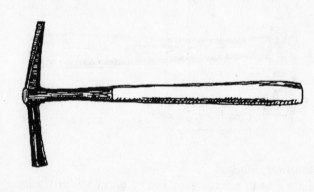

Polished steel head

• Hammer, Trimmer's

A trimmer's hammer is a lightweight hammer once commonly used by house trimmers—the finish-carpenters and joiners responsible for the fine interior woodwork of a house.

• Hammer, Upholsterer's

There are many varieties of upholsterer's hammers, but the hammer illustrated here was probably among the finest ever made. The head is made of extra-quality cast steel, the hickory handle is grooved for a better grip, and the tool was originally available with different size faces, from ³⁄₁₆ in. to ⁵⁄₈ in.

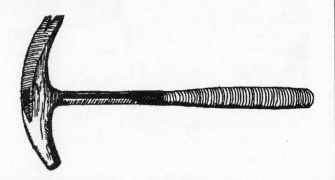

• Hammer, Veneer

The veneer hammer is not (by Nicholson's definition) strictly a hammer at all. It is used to press freshly glued veneer down by being moved across the surface from side to side.

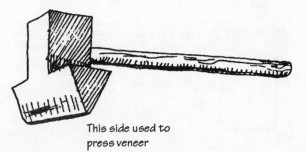

This side used to press veneer

• Hammer, Warrington Pattern

A relatively small cross-pein hammer, the Warrington pattern is the cabinetmaker's bench hammer of choice. Its cross pein is useful for starting small brads and nails, and its well balanced, light weight is ideal for the less violent blows generally required for fine furniture.

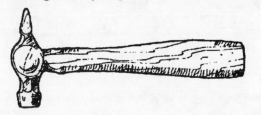

Hammer Holster

A hammer holster is worn on the belt and holds the carpenter's hammer securely and within easy reach.

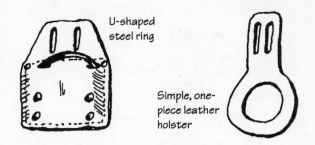

U-shaped steel ring

Simple, one-piece leather holster

Hand Drill: see *Drill (1)*.

Hand Miter: see *Miter Trimmer*.

Handsaw: see *Saw*.

Hand-screw: see *Clamp*.

Hand Vise: see *Vise*.

Hatchet

The word *hatchet* is the diminutive of the French word for *axe*: *hache*, and means simply a short-hafted (handled) axe. It has another distinctive characteristic, however, that makes it more than just a small axe; it is invariably provided with a hammer-like poll on the back of its head.

• Hatchet, Barrel

The barrel hatchet is a typical hatchet with a flat top and a nail-hammering poll. The blade is 2 ½ in. wide and the whole tool weighs 18 oz.

• Hatchet, Broad or Hewing

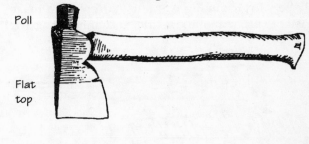

Poll

Flat top

A broad hatchet is a smaller version of the *broad axe*. It is not used to split wood nor to drive nails, but for shaping and dressing timbers. Like the broad axe, its cutting edge is beveled on one side only.

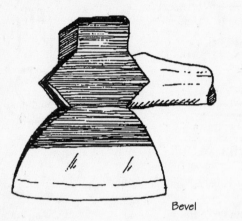

Bevel

• Hatchet, Claw

A claw hatchet is a medium-size hatchet, similar to a *broad* or *hewing hatchet*, but with the addition of a nail-pulling claw at the side of its nail-driving poll.

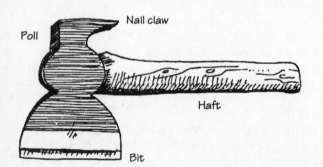

Nail claw

Poll

Haft

Bit

• Hatchet, Cooper's

This is the most unusually shaped hatchet, being used by coopers for trimming work. Its narrow head makes it easier to use within the restricted confines of a barrel.

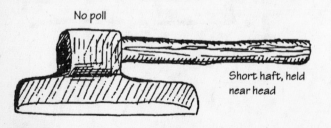

No poll

Short haft, held near head

• Hatchet, Half

A half hatchet is very similar to a *lath hatchet* in that the flare of the blade is confined to the side nearest the haft, thereby giving the impression that the blade has been cut in half.

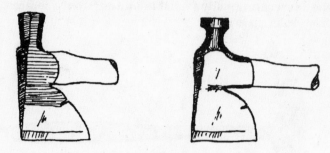

Two varieties of half hatchet

• Hatchet, Lath

Also called a "lathing hatchet," the lath hatchet has a flat top so that it may be used in corners or against ceilings when nailing lathing strips (which form the ground for plasterwork).

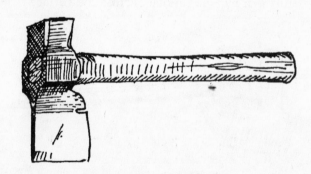

• Hatchet, Shingling

Probably the most common type still in use, the shingling hatchet is designed to be used for splitting and nailing wooden roof shingles and shakes (shingles that are sawn on one side and riven on the other).

Three shingling hatchets, all with nail-hammering polls and flared bits

Holdfast

Sometimes called a "bench holdfast," the holdfast is an L-shaped piece of iron used to hold a workpiece tightly to the benchtop. It is simply wedged into a hole bored in the benchtop. Modern versions are provided with an adjusting screw and a metal collar designed to prevent wear to the hole necessary in the benchtop.

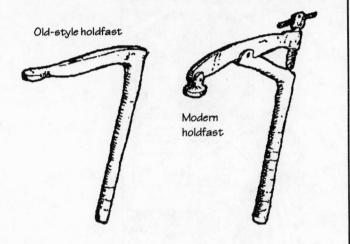

Old-style holdfast

Modern holdfast

Hollow Plane: see *Plane, Moulding*.

Hook Pin

Hook pins are used for drawing home the tenons of large beams into their mortises. After both parts of the mortise-and-tenon joint have been cut, a hole is bored through the side of the joint so that when assembled the hook pin may be hammered through, drawing the two parts tightly together. The hole in the tenon is slightly offset and closer to the shoulders of the joint than the hole bored through the cheeks of the mortise.

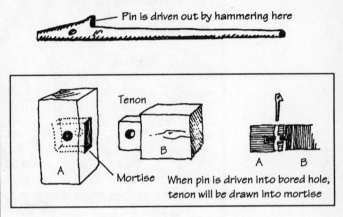

Pin is driven out by hammering here

Tenon

Mortise

A   B

When pin is driven into bored hole, tenon will be drawn into mortise

Horn Plane: see *Plane, Bench*.

The White King Watching the Carpenters, a woodcut by Weisskunig, ca.1515

Tools shown include: two-man crosscut saw (held by king), framing square
and broad axe (in use), trimming axe, chalk line, and sword saws.

Ice Axe: see *Axe*.

Improved Bevel: see *Bevel*.

Improved Miter Square: see *Square*.

Inclinometer

Inclinometers are instruments consisting of a blade that is placed against the work while a moveable *level*, mounted inside a *protractor*, is adjusted until its bubble indicates horizontality, allowing the angle of the work to the horizon to be read off the inscribed scale and transferred if desired.

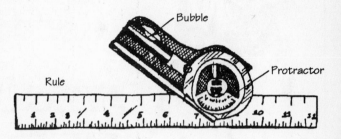

Inshave

The inshave is a single-handled form of *scorp*. Both these tools are related to *drawknives* and exist in a variety of forms used by coopers, shoemakers, and other specialized trades.

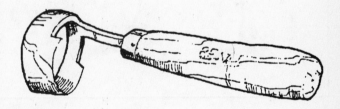

Inside Calipers: see *Calipers*.

Inside-start Saw: see *Saw*.

Iron Wedge: see *Wedge*.

Sawing Veneer, from *L'Art du Menuisier Ebéniste* (The Art of the Cabinetmaker), by André-Jacob Roubo, 1774

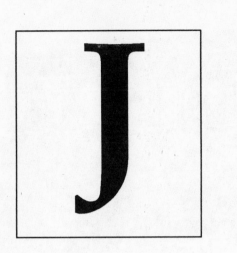

Jack Knife: see *Knife.*

Jack Plane: see *Plane, Bench.*

Jeweler's Hammer: see *Hammer.*

Jeweler's Mallet: see *Mallet.*

Jeweler's Saw: see *Saw.*

Jeweler's Screwdriver: see *Screwdriver.*

Jigger: see *Drawknife.*

Joiner Saw: see *Saw.*

Joiner's Hammer: see *Hammer.*

Jointer Plane: see *Plane, Bench.*

Joint Knife: see *Knife.*

Japanese Carpenters

Carpenters Preparing Timbers, from *L'Encyclopédie, ou Dictionnaire Raisonné des Sciences, des Arts et des Métiers* (The Encyclopedia, or Analytical Dictionary of the Sciences, the Arts and the Trades), by Denis Diderot, 1772

Tools shown include: framed pit saw, mallet, framing chisel, dividers, hewing axe, and twibill.

Keyhole Saw: *see Saw.*

Knife
While the knife, next to the *hammer*, is probably the most basic of all tools, and the one tool that almost everybody owns at one time or another, there are relatively few knives connected with woodworking compared to the number of non-woodworking knives, which range from table knives to hunting knives.

• Knife, Drawing
The old-time drawing knife (not to be confused with the *drawknife*) has been superseded by the *sloyd* or *bench knife*, which perform the same functions and which are made expressly for the purpose, the drawing knife being generally made from an old knife or discarded *chisel*. One of the drawing knife's particular jobs was to cut a V-shape groove along a line to be sawn, so that the saw blade would remain in the groove and not wander.

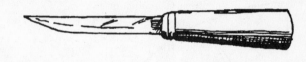

• Knife, Felt
The felt knife is used by piano tuners, cabinetmakers, and various other furnituremakers. Consequently, it exists in various patterns, although all are used to cut felt.

American felt   French   American   Tuner's
knife   trimming knife

Various patterns from 2"–6" in length

• Knife, Hacking

The hacking knife, used for general-purpose trimming much like a small *hatchet* or machete, is characterized by having a leather handle.

Leather handle

5" blade

• Knife, Joint

The joint knife is also called a "spackle" or "spackling knife," since it is used when applying spackle or joint compound to the joints between pieces of gypsum wallboard. It is a spreading tool rather than a cutting tool, related to the *putty knife*. There are many varieties with differently shaped blades to make the application of the compound easier in various situations, such as in corners.

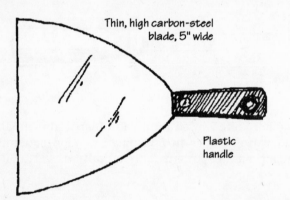

Thin, high carbon-steel blade, 5" wide

Plastic handle

• Knife, Paper-hanger's

The paper-hanger's knife is used by house decorators when hanging wallpaper.

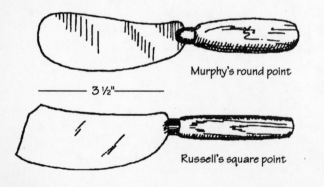

Murphy's round point

3 ½"

Russell's square point

• Knife, Plasterer's

Every old-time carpenter was often called upon to do a little plastering, and so would be likely to have a plasterer's knife in his toolbox.

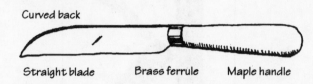

Curved back

Straight blade   Brass ferrule   Maple handle

• Knife, Pocket

No carpenter is without a pocket knife, of which there are innumerable varieties called variously "pen knives," "jack knives," and "clasp knives." Some have only one blade while others have many, but all share the characteristic of having blades that may be folded into the handle—thus making them pocket knives.

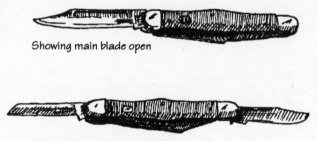

Showing main blade open

Showing two smaller blades open

• Knife, Putty

The putty knife, with its square and flexible blade intended for applying putty (nowadays replaced by glazing compound) to window panes, is also used for applying joint compound, making it, in effect, a narrower version of the *joint knife*.

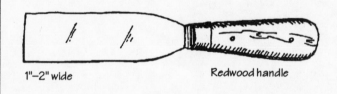

1"–2" wide   Redwood handle

• Knife, Razor Plow

The razor plow is an extremely efficient woodcarving knife designed by David Boyle. Held in the palm of the hand, it is pushed away from the worker with great power and control, and is especially well adapted for chip-carving and incised lettering.

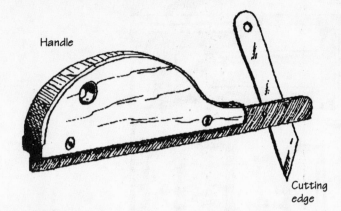

Handle

Cutting edge

• Knife, Sloyd

The sloyd knife is the woodworker's standard bench knife. One of its main uses is to mark joints and lay out joinery details.

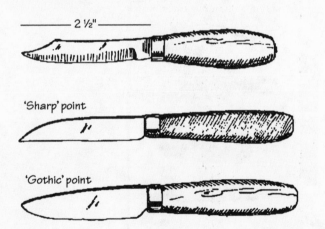

2 ½"

'Sharp' point

'Gothic' point

• Knife, Utility

The utility knife is an all-purpose knife used by framers, carpenters, and gypsum-wallboard installers. Its chief feature is that its blade (often retractable) is replaceable and disposable.

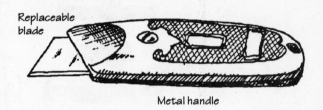

Replaceable blade

Metal handle

Knurling Tool

A knurl is a small wheel whose edge is so filed as to leave a particular impression when pressed into the work. Knurling tools are usually made with a set of interchangeable knurls or wheels designed to fit into an adjustable holder.

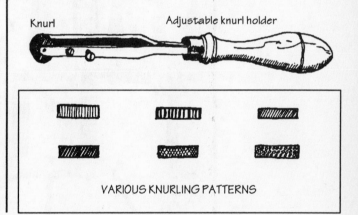

Knurl

Adjustable knurl holder

VARIOUS KNURLING PATTERNS

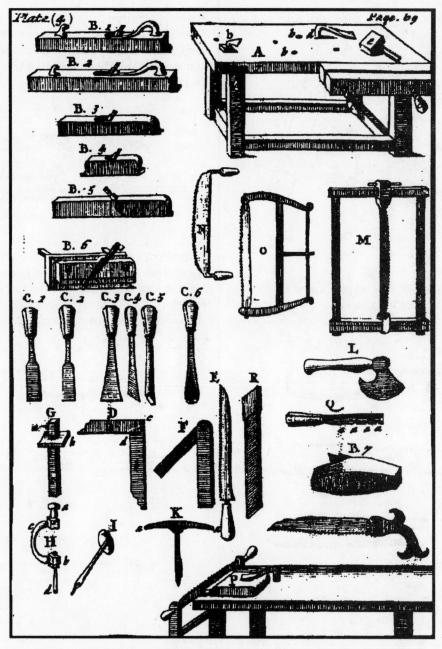

Joyners Tools, from *Mechanick Exercises or the Doctrine of Handy-Works,*
by Joseph Moxon, 1703

Tools shown include: A. workbench, B. fore plane, B2. jointer plane, B3. Jack plane, B4. and B7. smooth plane,
B5. long rabbet plane, B6. plow plane, C1 and C2. firmer chisels, C3. paring chisel, C4. skew chisel, C5. mortise chisel,
C. gouge, D. trysquare, E. compass saw, F. bevel, G. marking gauge, H. bit brace, I. gimlet, K. auger, L. hatchet, M. pit saw,
N. whipsaw, O. frame saw, P. holdfast, Q. saw wrest, R. file.

Last Maker's Rasp: *see Rasp.*

Lath Hammer: see *Hatchet.*

Lath Hatchet: see *Hatchet.*

Level

A level is an instrument for determining whether something is perfectly horizontal. There are two types of level: the older A-level, known by the Egyptians and used for centuries, and which is still in use today in many parts of the world, and the spirit level, which, although the only level most people now know, took almost two hundred years to gain acceptance after its invention in France in 1666.

The A-level consists of a triangular frame, shaped like the letter *A*, from the apex of which is suspended a *plumb bob.* When the level is positioned so that the plumb bob is centered over the bottom member of the frame, the surface on which the level is resting must be horizontal.

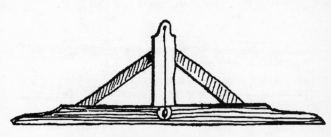

19th-century A-level

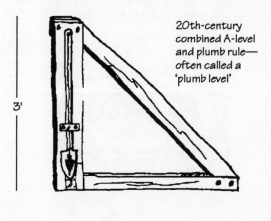

3'

20th-century combined A-level and plumb rule— often called a 'plumb level'

The spirit level consists of a slightly curved, clear tube, nearly filled with liquid, usually alcohol and often colored, set in a wooden or metallic beam in such a way that when the beam is held horizontally, the bubble of air remaining in the tube will be centered at the top of the curve, usually between two marks. The spirit level is so called because its inventor, the Frenchman Thevenove, used wine in the tube in order that it might be more easily seen and so that it would not freeze as quickly as water.

Bubble

When bubble is centered in vial, level is horizontal

• Level, Adjustable (Adjustable Plumb, Level, and Inclinometer)

Also known as an adjustable plumb, level, and inclinometer, this tool may be used to establish any degree from the horizontal to the vertical by rotating the central vial and reading the scale, calibrated to a full 360°.

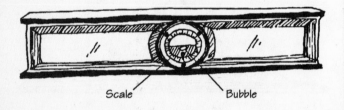

Scale   Bubble

• Level, Bull's-eye

The bull's-eye level is a small, round, pocket level used to check quickly the horizontality of various surfaces, such as table tops and machine surfaces.

Bubble

• Level, Cross-test

The cross-test level has two bubbles united in a single frame so that when placed on work to be leveled in two directions it is not necessary to move the tool.

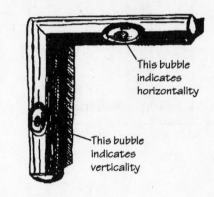

This bubble indicates horizontality

This bubble indicates verticality

• Level, Digital

Many varieties of level are now provided with digital readouts instead of a directly observable bubble. Providing the tool is accurately calibrated and consistently powered (two attributes not always easily confirmable), a great degree of accuracy is possible.

• Level, Line

Line levels are very small levels about 3 in. long and little thicker than a pencil, that may be hung on a line stretched between two points. Provided the level is placed at the center of the line, which must be stretched as tautly as possible, a centered bubble will indicate that the two points are level. A typical use is for ensuring that opposite foundation walls are raised to the same height.

Level is hung on taut line from hooks at each end

Bubble

• Level, Machinist's

A machinist's level is an extremely accutrate level with bottoms precision-milled to ensure absolute flatness, and which is further fitted with fine-adjusting screws on the tube.

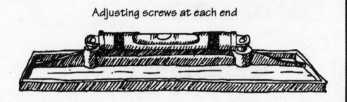

Adjusting screws at each end

• Level, Mason's
A typical mason's level is 4 ft. long, and usually combines vertical bubbles with several horizontal bubbles. If made of wood, the stock is frequently metal-bound to resist wear.

—————— 4' ——————

• Level, Pocket
Although pocket levels are small (rarely more than 2 in. to 4 in. long), they are made in a variety of shapes.

Vial in metal tube

Vial in glass tube

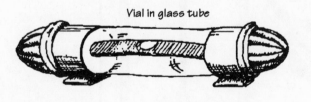

• Level, Spirit (Plumb Level)
The development of the spirit level is described under *level*. The fact that the same principles are employed when including an additional spirit-filled tube at the end of the beam or stock (thereby turning the tool into an instrument that can measure verticality as well as horizontality) effectively turns today's standard levels into *plumb rules* as well. Although a contradiction in

terms, such tools are often known as "plumb levels." The average length is 2 ft. They may be made with a wooden stock, in which case a straight-grained hardwood unlikely to warp, such as cherry or mahogany, is preferred, or with a metallic stock, usually of aluminum, which is light but strong.

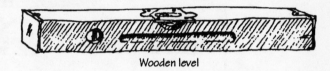

Wooden level

Aluminum level

• Level, Torpedo
The torpedo level is a short level, a little bigger than a *pocket level*, made in the shape of a torpedo. Like other modern levels, it may be made of wood or metal.

Vertical bubble   Horizontal bubble   45° bubble

Level Sights
By attaching a pair of level sights to a *spirit level*, the user has a convenient and accurate way of leveling from one given point to another, some distance away. In effect, the sights transform a spirit level into a form of hand-held transit—a surveyor's leveling instrument usually mounted on a tripod.

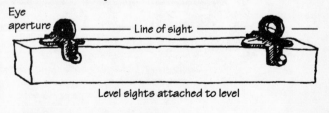

Eye aperture   —— Line of sight ——

Level sights attached to level

Long Plane: see *Plane, Bench.*

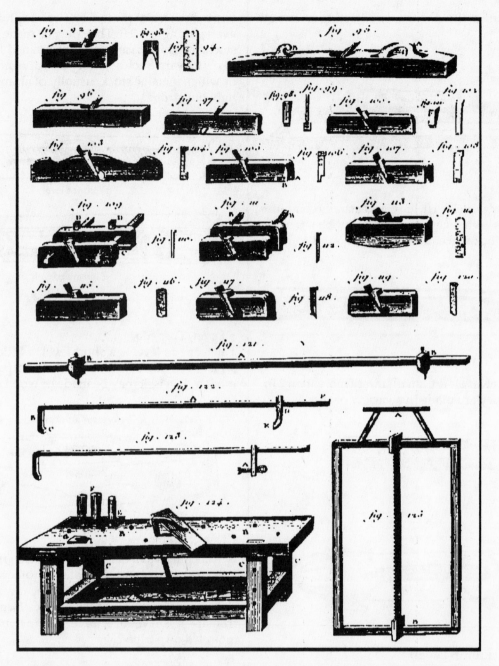

Tools of the Joiner, from *L'Encyclopédie, ou Dictionnaire Raisonné des Sciences, des Arts et des Métiers*
(The Encyclopedia, or Analytical Dictionary of the Sciences, the Arts and the Trades), by Denis Diderot, 1769

Tools shown include: smooth plane, jointer plane, jack plane, rabbet plane, plow plane,
compass plane, tonguing plane, grooving plane, moulding planes, trammel points,
clamps, frame saw, bench, holdfast, chisels, and bench dog.

Macaroni: see *Chisel, Woodcarving, Box Tool.*

Magazine Brad Awl: see *Awl.*

Mallet
A mallet may be described as a kind of wooden *hammer*. The word *mallet* actually means a small *maul* (which word derives from the Latin word for hammer: *malleus*!) The woodworker generally uses a mallet for striking wood, such as wooden-handled chisels, and a *hammer* only for striking metal things, such as nails or *cold chisels*.

• **Mallet, Carver's**
The distinctive feature of the woodcarver's mallet is its round head, which enables a chisel to be struck from any angle. Carver's mallets are usually made from the very hardest woods, such as dogwood or lignum vitae.

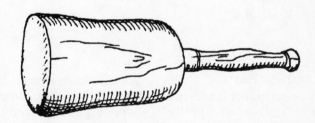

• **Mallet, Fiber-head**
The heads of these mallets are made of vulcanized fiber, which is very hard and dense, extremely tough, and yet elastic enough to form a cushion so that the hand is not jarred by forceful blows. Because there are malleable iron rings around the head, this mallet is also called a "ring mallet." For greater security, the handle is screwed rather than wedged into the head.

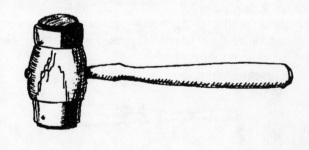

• Mallet, Froe Club

The froe club is the particular mallet used for striking *froes*, and is usually user-made from any convenient billet of hardwood.

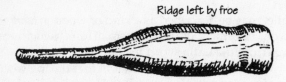

Ridge left by froe

• Mallet, Jeweler's

The jeweler's mallet is the smallest and lightest of the mallets. Although the best quality mallets are made of lignum vitæ, hickory is sometimes used.

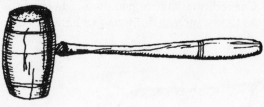

2" face

• Mallet, Oval

The oval mallet is a form of bench mallet used by those woodworkers who prefer its shape and balance to the other two common forms: the *round mallet*, and the *square mallet*.

3 ½" face

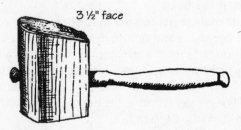

• Mallet, Rawhide

The rawhide mallet is a lightweight mallet made, except for its handle, entirely of hide. It is used when the work being hit must not be marred.

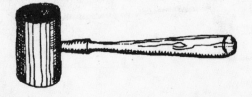

• Mallet, Ring

The ring mallet is intended for fairly heavy work, and is therefore provided with iron rings or bands that prevent the faces from splitting or mushrooming as a result of continued pounding.

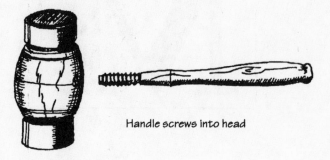

Handle screws into head

• Mallet, Round

The round mallet is one of the most common mallets used by carpenters. The head is usually made of turned hickory or lignum vitæ, and its handle is wedged into the head.

2"–4" face

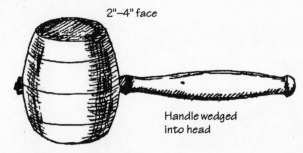

Handle wedged into head

• Mallet, Square

The square mallet is the woodworker's standard bench mallet. British mallets are often made of beech; American mallets are often made of maple.

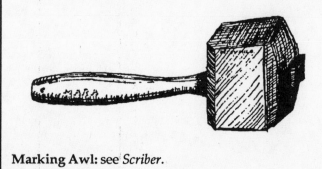

Marking Awl: see *Scriber.*

Marking Gauge: see *Gauge.*

Mast Plane: see *Plane, Special-purpose.*

Match Planes: see *Plane, Special-purpose.*

Maul

The word *maul* comes from the Latin word: *malleus,* meaning *hammer.* The maul, however, was not originally made of iron like a *hammer,* but of wood. It is, in fact, a large *mallet*—the word *mallet* being the diminutive of maul.

• Maul, Beetle

Beetle is used synonymously with maul to refer to the standard, long-handled maul. *Beetle* derives from the Old English word: *beatan,* meaning to beat—it has nothing to do with the insect (whose name comes from another word meaning "to bite"). In fact, a beetle is a wooden *sledge hammer,* used for knocking heavy wooden beams into place, ramming paving stones, and driving *splitting wedges* into logs.

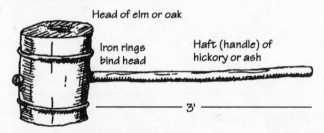

Head of elm or oak

Iron rings bind head

Haft (handle) of hickory or ash

3'

• Maul, Five-pound

The tool that is generally called by house framers a "five-pound maul," although used for the same purpose as old wooden mauls (i.e., knocking posts and beams and sections of framing into place), is in reality a heavy *hammer,* since its head is made of iron and not of wood.

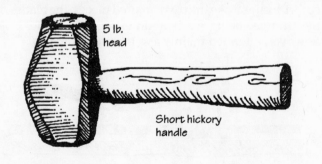

5 lb. head

Short hickory handle

In fact, its nickname, "Thor's hammer," is more accurate than the name it is sold under, despite the fact that Thor (the Norse god of thunder) probably had a hammer of stone.

• Maul, Rawhide

A rawhide maul is a curious combination of *hammer* and *mallet,* being made of wood and iron covered with rawhide. There are also *rawhide hammers* as well as *rawhide mallets;* the rawhide maul is heaviest of all.

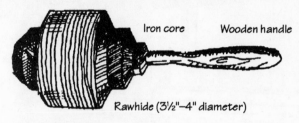

Iron core   Wooden handle

Rawhide (3½"–4" diameter)

Miter Block: see *Miter Box.*

Miter Box (Br. Mitre Box)

A miter box is a device for guiding a *saw* at the required angle. In its simplest (and original) form it is intended to guide the saw through the work so that the piece sawn forms part of a common (45°) miter joint (*see box on following page*).

• Miter Box, Folding

The folding miter box's advantage is that it takes up very little space in a toolbox. The base is comprised of two leaves that may be folded up against the vertical central piece when not in use. A sliding pin locks them in place when opened.

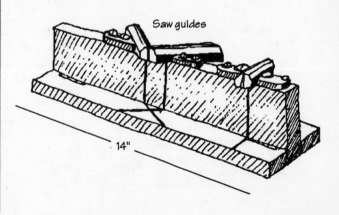

Saw guides

14"

• Miter Box, Metal

A metal miter box is a very sophisticated device containing such refinements as special *clamps* to hold the work in place, catches to hold the *saw*, and special marks to facilitate cutting particular shapes. Together with metal *chute boards*, metal miter boxes are among the most expensive woodworking handtool devices ever made.

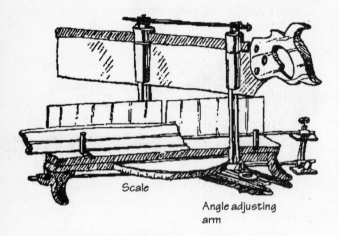

Scale

Angle adjusting arm

• Miter Box, Miter Block

At the opposite extreme from the *metal miter box*, the miter block—the simplest and least expensive miter box—consists merely of a simple block of wood in which a rabbet or shelf has been formed (or of two pieces of wood joined to form a similar shelf), with saw cuts in the upper section made to guide the *saw* while the workpiece rests on the lower part.

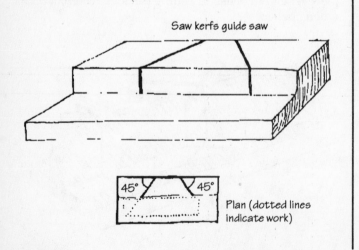

Saw kerfs guide saw

45°   45°

Plan (dotted lines indicate work)

• Miter Box, Wooden

More accurate than the *miter block* and less expensive than the *metal miter box*, the wooden miter box is a nice compromise. It is probably the most common form of the tool, and is often user-made to suit the occasion. Factory-made boxes are invariably made of maple, since this is a very stable and hard-wearing wood.

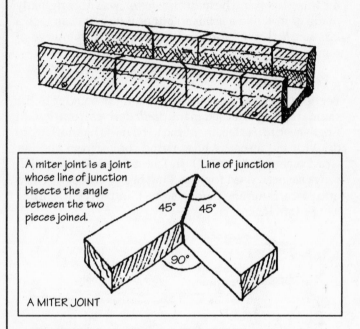

A miter joint is a joint whose line of junction bisects the angle between the two pieces joined.

Line of junction

45°   45°

90°

A MITER JOINT

Miter-box Saw: see *Saw*.

Miter Plane: see *Plane, Special-purpose*.

Miter Trimmer (Br. Mitre Trimmer)

Also known as a "hand miter," the miter trimmer works like a paper guillotine. It quickly and accurately trims the 45° faces of wood prepared for mitering. Small pieces (such as moulding and picture framing) can actually be cut with one stroke into two mitered pieces, the blades of the trimmer being made at an exact right angle. It is, however, intended mainly as a trimming device.

Knives

Mitre Shooting Block: see *Clamp, Miter-planing*.

Mortise Chisel: see *Chisel, Bench*.

Mortise Gauge: see *Gauge*.

Moulding Plane: see *Plane, Moulding*.

Moulding Tool
The moulding tool, used in carriage-building, is a cross between a *moulding plane*, and a *spokeshave*, while at the same time being closely related to the *beading tool*. Used for forming moulded profiles on curved sections of wood, it is made in various patterns, but always with a left and a right side so that it may accommodate grain direction in the most felicitous fashion.

Moving Filletster: see *Plane, Special-purpose, Rabbet*.

The Woodworker's Shop, by Jan Joris van Vliet, 17th century

Tools shown include: fore plane, smooth plane, square, auger, axe, brace, chisel, sword saw, adze, dividers, and broad axe (in use).

Nail Claw

A nail claw, which is very often called a "cat's paw," is used for removing nails. Using a *hammer*, its claw is beaten under the head of a nail already driven into wood. By levering the tool backwards, again with the help of hammer blows to the other end of the handle, the nail may be pried out.

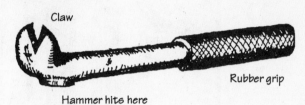

Claw

Rubber grip

Hammer hits here

Nail Puller

Like the *nail claw*, the nail puller is also used for removing nails. The jaws are positioned on either side of the nail's head, the shank is drawn out to its fullest extension, and then rammed back onto the foot, causing the jaws to dig into the wood. When the puller is levered backwards, the hinged jaw digs further into the wood and the nail is pried out (*see Frontispiece*).

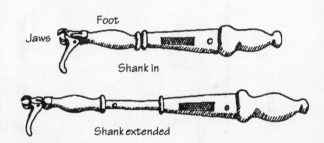

Foot

Jaws

Shank in

Shank extended

Nail Set

A nail set is a metal punch used by woodworkers for "setting" nails below the surface of the wood into which they have been driven. By using the set, the surrounding wood remains unblemished by *hammer* strokes. The points of nail sets are made concave in order to avoid slipping off the nail. Four sizes are common: 1/32 in., 1/16 in., 3/32 in., and 1/8 in.

Nosing Plane: see *Plane, Moulding*.

Notching Chisel: see *Chisel, Wood, Bench*.

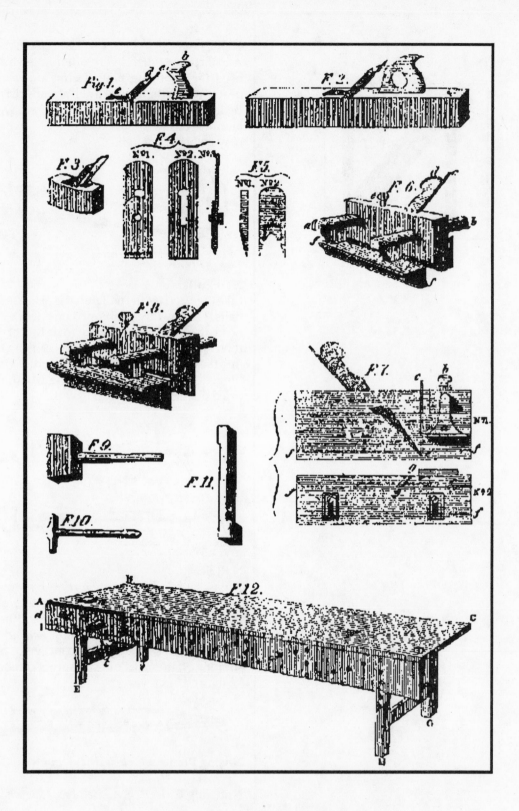

Offset Screwdriver: see *Screwdriver*.

Ogee Plane: see *Plane, Moulding*.

Oilstone: see *Sharpening Stones*.

Old Woman's Tooth: see *Plane, Special-purpose*.

Outside Calipers: see *Calipers*.

Oval Mallet: see *Mallet*.

Oval Scraper: see *Scraper*.

Ovolo Plane: see *Plane, Moulding*.

facing page:
Joinery Tools, from *Mechanic Exercises*, by Peter Nicholson, 1812

Tools shown include: jack plane, trying plane, smooth plane,
plow plane, mallet, hammer, bench dog, filletster, and bench.

The Cabinetmaker's Shop, from *Orbus Pictus nach Zeichnungen der Susanna Maria Sandrart*
(The World in Pictures from Drawings by Susanna Maria Sandrart), by Elias Pozelius, 1690

Tools shown include: chisels, glue pot, gimlet, jack plane, and frame saw.

Panel Gauge: see *Gauge.*

Panel Saw: see *Saw.*

Paper-hanger's Knife: see *Knife.*

Paring Chisel: see *Chisel, Wood, Bench.*

Paring Gouge: see *Chisel, Wood, Gouges.*

Parting Tool: see *Chisel, Wood, Turning.*

Patternmaker's Chisel: see *Chisel, Wood, Bench.*

Patternmaker's Gouge: see *Chisel, Wood, Gouges.*

Patternmaker's Router Plane: see *Plane, Special-purpose.*

Patternmaker's Shrinkage Rule: see *Rule.*

Peeling Chisel: see *Chisel, Wood, Special-purpose.*

Peg Awl: see *Awl.*

Pen Knife: see *Knife.*

Pianomaker's Edge Plane: see *Plane, Special-purpose.*

Pianomaker's Plane: see *Plane, Special-purpose.*

Pincers
A pair of pincers is a holding tool similar to a pair of *pliers,* but, as the name implies, made with sharp jaws that pinch and may be used to cut wire or nails. Pincers are most usually used by the woodworker to pull out small nails and tacks.

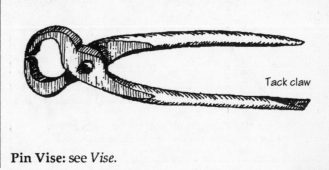

Tack claw

Pin Vise: see *Vise.*

Pitch Adjuster

A pitch adjuster is a device that may be attached to a *level* in order to determine the pitch of a particular slope. This is commonly stated in terms of how much the slope rises (or falls) in inches per running foot. For this reason the scale is graduated in inches. (Of course, if the level to which the pitch adjuster is attached is longer than 1 ft., the pitch is stated as being so many inches per the length of the level being used.)

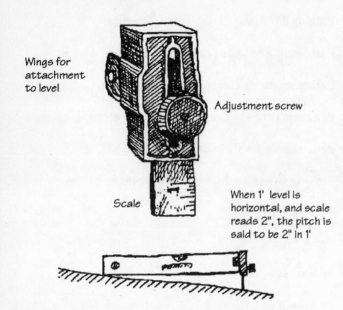

Wings for attachment to level

Adjustment screw

Scale

When 1' level is horizontal, and scale reads 2", the pitch is said to be 2" in 1'

Pit Saw: see *Saw.*

Plane

The word *plane* comes from the Latin word: *planus,* meaning "level." Hence a plane is a tool used for leveling (and smoothing) the surface of wood by paring shavings from it. The first planes were certainly no more than a *chisel* held in a block of wood to control the depth of cut more easily.

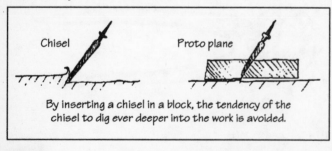

Chisel   Proto plane

By inserting a chisel in a block, the tendency of the chisel to dig ever deeper into the work is avoided.

A plane consists essentially of a stock or body (of wood or metal) in which a blade (the iron) is set. The iron projects slightly through a slit called the mouth, cut in the sole (the underside) of the plane's stock.

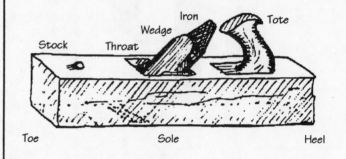

Stock   Throat   Wedge   Iron   Tote

Toe   Sole   Heel

Since planes are undoubtedly more diversified than any other woodworking tool, they have been divided into the following four categories:

1. **Bench Planes**
2. **Block Planes**
3. **Special-purpose Planes**
4. **Moulding Planes**

1. Bench Planes

Bench planes are used for reducing, leveling, and smoothing wood. They are listed in order of size (which is not necessarily the order in which they may be used), each size being dedicated to a different purpose. Over the years, while the relative size of these planes has remained the same, individual measurements have changed somewhat; the sizes indicated in the following table represent the average limits of each tool.

Smooth Plane	5 ½ in. to 10 in.
Horn Plane	5 ½ in. to 20 in.
Jack Plane	12 in. to 17 in.
Fore Plane	18 in.
Trying Plane	20 in. to 22 in.
Long Plane	26 in.
Jointer Plane	22 in. to 36 in.

Average Sizes of the Various Bench Planes

It should be realized that originally all planes were made with wooden stocks. Until the middle of the eighteenth century there were a wide variety of designs;

thereafter, as planemaking became a separate trade, designs were relatively standardized following the model of those tools produced by the English toolmakers in Sheffield. By the end of the nineteenth century, however, American metal-bodied planes had become common, and wooden-bodied planes are now made in only a very few places. All the following bench planes exist in both wooden and metallic versions.

An interesting exception are those high-quality cabinet-makers' planes made around the turn of the last century in Britain known as "stuffed planes." Like the planes used centuries before by the Romans (*see box on page 95*), these planes were made with metal bodies "stuffed" or infilled with hardwood. They represent the acme of planemaking, and are eagerly sought after by collector and user alike.

• Smooth Plane

The smooth or "smoothing" plane is the last plane used in giving the utmost degree of smoothness to the surface of the wood. It is used chiefly in cleaning off finished work. For this reason it is the smallest of the bench planes.

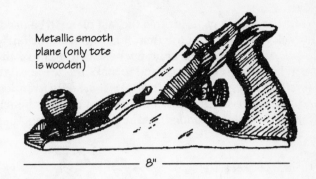
Metallic smooth plane (only tote is wooden)

8"

• Horn Plane

The presence of the horn puts this variety of wooden-bodied plane, originally native to Germany, in a separate class. It is a curious anomaly in the plane world, being the only continental European plane to have become common in the British-American toolbox. It is made in a variety of sizes embracing *smooth planes*, *jack planes*, and *try planes*.

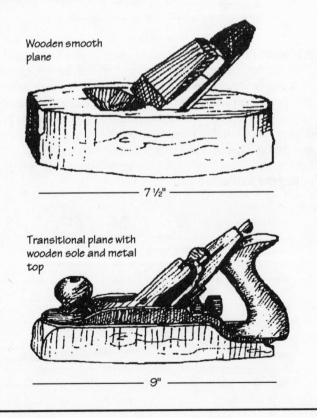
Wooden smooth plane

7 ½"

Transitional plane with wooden sole and metal top

9"

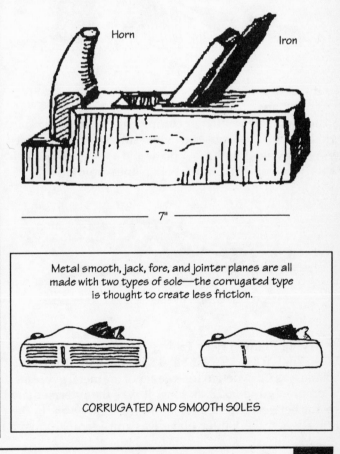
Horn   Iron

7"

Metal smooth, jack, fore, and jointer planes are all made with two types of sole—the corrugated type is thought to create less friction.

CORRUGATED AND SMOOTH SOLES

• Jack Plane

While the *smooth plane* is often the last plane to be used, the jack plane is often the first. It is intended for fairly heavy, coarse work. Its chief purpose is to remove the irregularities left by the *saw* and to make the surface flat. The cutting edge of its iron (blade) is markedly rounded and more grossly protuberant than that of other bench planes, since it is intended to take bigger shavings.

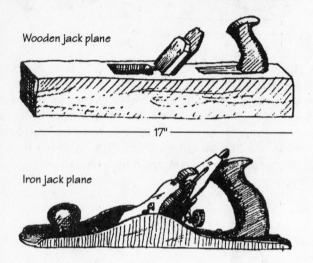

Wooden jack plane

17"

Iron jack plane

• Fore Plane

A fore plane has a flatter cutting edge than a *jack plane*, and is intended for use on longer surfaces. It may sometimes double as both a jack plane and a *jointer plane* when it is inconvenient to carry both planes around.

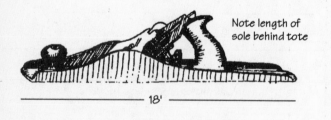

Note length of sole behind tote

18'

• Trying Plane (Try or Truing Plane)

Also known as a try plane and a truing plane, the trying plane was the wooden forerunner of the metal *fore plane*. Commonly made 22 in. long, it was the intermediate plane between the *jack plane* and the *jointer plane* in the days when the jointer plane was considerably longer than it is today. In Britain the term *try plane* usually means a jointer plane.

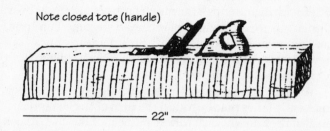

Note closed tote (handle)

22"

• Long Plane

The long plane is really a short *jointer plane*, although it is, in fact, longer than most modern jointer planes. A hundred years ago, when jointer planes were often longer than they are now, the long plane was the intermediate plane between the *trying plane* and the jointer plane.

26"

• Jointer Plane

The jointer plane is the longest of all bench planes, ranging from 2 ft. to 3 ft. Because of its extreme length, when used on a long board it will cut at first only the high spots, progressively lengthening the surfaces thus produced, until finally the entire edge being planed lies in one plane. This process is known as "jointing"' and is the necessary prerequisite for joining two boards together—hence the tool's name.

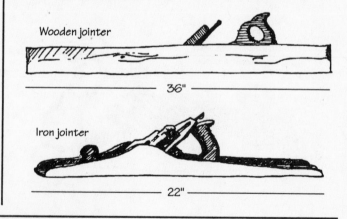

Wooden jointer

36"

Iron jointer

22"

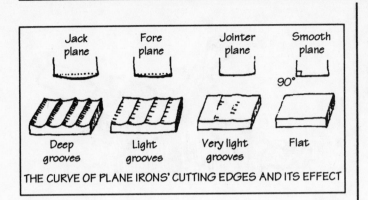

Jack plane	Fore plane	Jointer plane	Smooth plane

Deep grooves / Light grooves / Very light grooves / Flat

THE CURVE OF PLANE IRONS' CUTTING EDGES AND ITS EFFECT

2. Block Planes

The block plane is a small, one-handed plane used for light work and especially for planing end-grain, for which purpose its iron (blade) is set at a much lower angle than that of the *bench planes*. Block planes developed from *miter planes* that were used for planing the end-grain of butcher block (hence the name), and eventually became the dominant form, virtually ousting the miter plane from the contemporary toolbox.

By the end of the last century, tool manufacturers had produced an astonishing number of varieties of this simple plane; most differ only in relatively inessential details. The following list compromises three distinct types:

> Iron Block Plane
> Block and Rabbet Plane
> Cabinetmaker's Block Plane

• Iron Block Plane

The iron block plane illustrated here represents the basic type, of which very many variations exist. Although all are approximately the same size, the differences consist of slight variations in iron (blade) angle and the addition of features such as adjustable mouths and different iron-holding arrangements. The common iron block plane serves as a handy, all-purpose tool for light work, as well as for plaining end-grain.

7"

• Block and Rabbet Plane

The block and rabbet plane is an example of a more complicated form of block plane. Its iron (blade) is skewed, one of its sides may be removed, effectively transforming it into a *rabbet plane*, and it is fitted with a fine depth-adjustment mechanism for the iron.

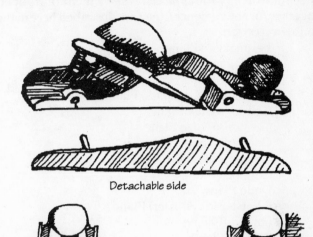

Detachable side

Used as block plane   Used as rabbet plane

• Cabinetmaker's Block Plane

This particular two-handed plane is actually closer to the *miter plane* from which it is descended, and is intended for extremely high-quality work. Its sides are square, allowing it to be used on its side for square planing, and its mouth is adjustable to a very narrow opening.

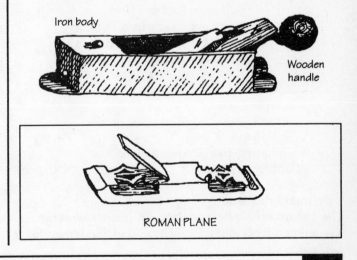

Iron body

Wooden handle

ROMAN PLANE

3. Special-purpose Planes

While *bench planes* and *block planes* are used for basic stock preparation (flattening and smoothing), and *moulding planes* are used for forming profiles on the edges of wood, special-purpose planes are used for all the other numerous operations possible with planes. With the exception of the various *rabbet planes* (which are grouped together for convenience), they are described here in the following order:

>   Beltmaker's Plane
>   Carriagemaker's Plane
>   Chamfer Plane
>   Circular (or Compass) Plane
>   Cooper's Chiv Plane
>   Cooper's Croze Plane
>   Cooper's Long Jointer
>   Cooper's Sun Plane
>   Core Box Plane
>   Dado Plane
>   Filletster (or Fillister) Plane
>   Forkstaff Plane
>   Furring Plane
>   Gutter Plane
>   Mast (or Spar) Plane
>   Match Planes
>   Miter Plane
>   Patternmaker's Sole Plane
>   Patternmaker's Plane
>   Patternmaker's Router
>   Pianomaker's Edge Plane
>   Plow Plane
>   Rabbet Plane
>   Adjustable Rabbet Plane
>   Bull-nose Rabbet Plane
>   Carriagemaker's Iron T-plane
>   Carriagemaker's Rabbet Plane
>   Filletster Plane
>   Side Rabbet Plane
>   Router Plane (Old Woman's Tooth)
>   Scraper Plane
>   Scrub Plane
>   Spill Plane
>   Tonguing and Grooving Plane
>   Universal Plane

• Beltmaker's Plane

The beltmaker's plane is used by leather-workers to shave the ends of belting so that it may be joined smoothly.

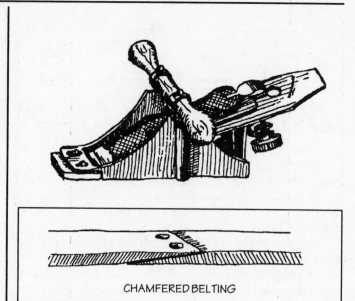

CHAMFERED BELTING

• Chamfer Plane

The sole of the chamfer plane illustrated here is V-shaped in order to ride on the arris of a square-edged piece of wood. The front part of the plane containing the iron (blade) is adjustable, enabling chamfers (45° bevels) of various widths to be formed. Many other varieties of chamfer plane exist, most of them wooden, some being very similar to *moulding planes*.

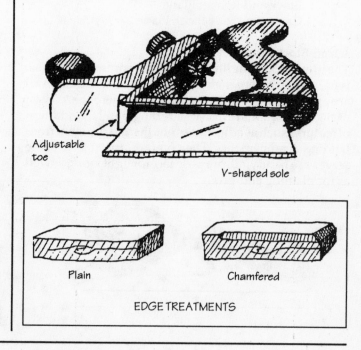

Adjustable toe

V-shaped sole

Plain   Chamfered

EDGE TREATMENTS

• Circular Plane (Compass Plane)

A circular plane, also known as a "compass plane," is designed to plane curved surfaces. Early wooden circular planes are able to plane only convex surfaces, and many of these planes were designed for one specific curve. Later wooden models were adjustable, but only the metal variety is capable of planing convex *and* concave surfaces by virtue of a flexible steel sole that may be adjusted to the desired curve.

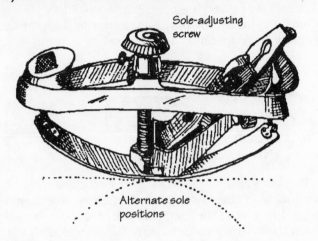

• Cooper's Chiv Plane

Chiv is the British name for the concave surface, known as the howel in America, cut in the inside ends of barrel staves to accommodate the groove in which the barrel heads will be fixed. The chiv plane therefore has a sole that is rounded in two directions and a large fence designed to ride on the ends of the staves.

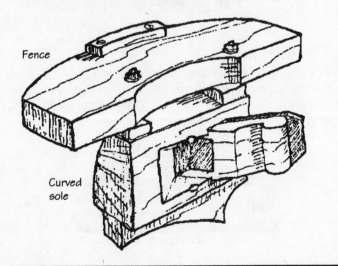

• Cooper's Croze Plane

The croze is a curious tool, half plane and half large *panel gauge*, designed to cut the groove in the howel or chiv (*see previous entry*) inside barrels in which the barrel head fits.

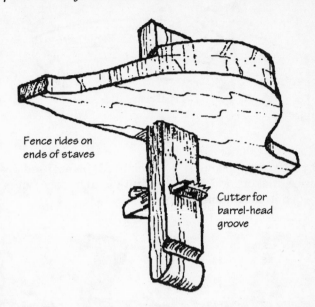

• Cooper's Long Jointer

The cooper's long jointer is the largest of all planes, often attaining a length of 6 ft. Because of its extreme size it is supported upside down, and the work is brought to the plane rather than vice-versa. It is used primarily for planing the sides of barrel staves.

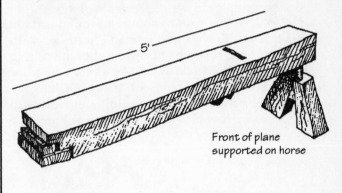

• Cooper's Sun Plane

The sun plane, made both left and right-handed, is a laterally curved plane designed to level the ends of barrels after the staves have been assembled and beveled. This leveling and concomitant forming of a small flat surface is necessary in order to provide a bearing surface for the *chiv plane* and *croze plane*.

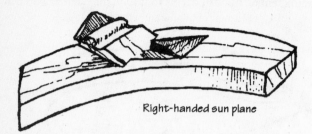

Right-handed sun plane

• Core Box Plane

The core box plane is a specialized plane used by patternmakers to excavate semi-circular concavities needed in the making of moulds.

Cross-section

V-shaped sole
Iron

• Dado Plane

A dado is a groove cut across the grain of a board. In order to achieve this with minimum tearout, the dado plane is made with a skewed iron and furnished with spurs designed to sever the fibers ahead of the cutting edge. Originally made of wood like a *moulding plane* (with which it is often confused), metal versions are common. Both varieties were made in different widths to cut dadoes as narrow as ¼ in. and as wide as 1 in.

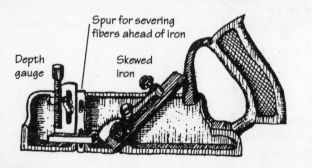

Spur for severing fibers ahead of iron

Depth gauge

Skewed iron

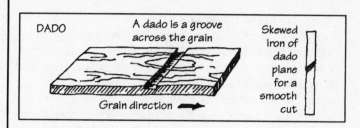

DADO   A dado is a groove across the grain   Skewed iron of dado plane for a smooth cut

Grain direction

• Forkstaff Plane

The forkstaff plane is a benchplane-sized tool used for planing cylindrical work. It is similar in design to the slightly more specialized *mast plane*.

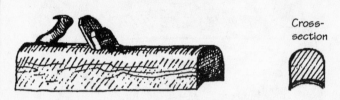

Cross-section

• Furring Plane

The furring plane is a nineteenth-century invention designed to remove the rough surface of wood as it comes from the sawmill, prior to its being planed with a *jack plane*. Its unusual sole makes removing the furry fibrous material and grit of undressed lumber relatively easy.

Recessed sole in front of and behind cutting iron

• Gutter Plane

The gutter plane is the reverse of the *forkstaff plane*, having a convex sole, designed, among other things, for planing gutters. A "gutter" in woodworking may refer to any long, concave surface.

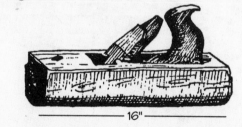

— 16" —

• Mast Plane (Spar Plane)

The mast, or spar, plane is a shipbuilding tool used in the finishing of cylindrical wooden masts and spars. Similar in size to a *jack plane*, the top part of the stock is shaped like a *smooth plane* to make it less bulky to hold.

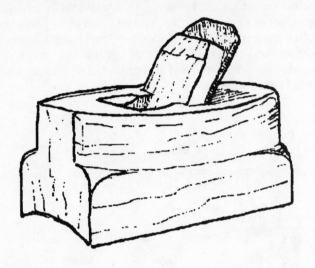

• Match Planes

The term "match planes" indicates a pair of planes used to cut matching parts, such as the tongue and the groove of a tongue-and-groove joint (*see box on page 103*), or some other two-part joint. Very often similar in form to *moulding planes* in appearance, these planes nevertheless are definitely not moulding planes since they cut no ornamental profile but parts of a joint.

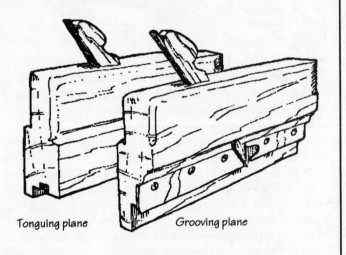

Tonguing plane   Grooving plane

• Miter Plane (Br. Mitre Plane)

Once no more than a simple block of wood with a low-angle iron (blade) designed to plane end-grain (and often referred to therefore as a "block plane"—not to be confused with its descendant, the *block plane*, so-named because of its use in planing butcher block), the miter plane evolved into a metal-bodied plane with wood infill, capable of extremely fine work on all sorts of grain.

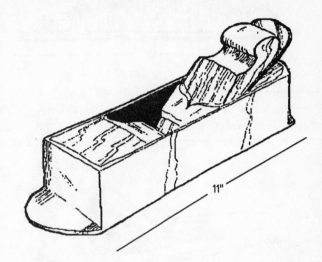

11"

• Patternmaker's Router Plane

Similar to a regular *router plane*, the patternmaker's router has a larger sole that is more carefully ground for the more exacting work demanded of it.

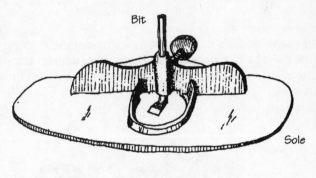

Bit

Sole

• Patternmaker's Sole Plane

The patternmaker's sole plane is a sophisticated version of the *gutter plane*, being designed to plane long, concave sections. Whereas the gutter plane can only plane a concavity of a fixed diameter, the patternmaker's sole plane is commonly provided with a set of differently sized, detachable soles, and furnished with a set of irons (blades) to match.

Curved, detachable beech sole

13"

• Pianomaker's Plane

Pianomaker's planes are small wooden planes, very similar to wooden *smooth planes*, but with their irons (blades) set at a lower angle, much like metal *block planes*.

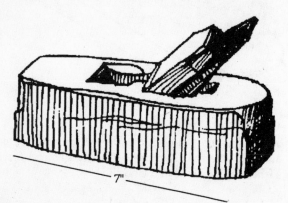

7"

• Pianomaker's Edge Plane

The pianomaker's edge plane is a low-angle finishing plane used by pianomakers and cabinetmakers for trimming inside work in places where other planes are unable to fit.

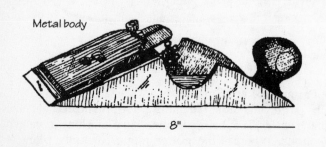

Metal body

8"

• Plow Plane (Br. Plough Plane)

A plow plane is designed to cut a groove, with the grain, some distance from the edge of the workpiece. Its fence regulates the position of the groove relative to the edge, its depth stop controls the depth of the groove, and a selection of different irons (blades) makes possible grooves of various widths. The wooden plow plane was often the centerpiece of a woodworker's toolkit, being an ornate and expensive piece of equipment, frequently made of exotic hardwood and provided with many brass and ivory fitments.

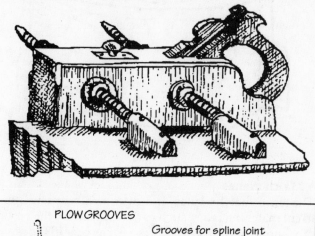

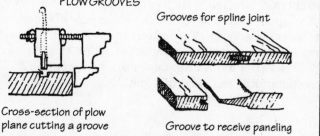

PLOW GROOVES

Grooves for spline joint

Cross-section of plow plane cutting a groove

Groove to receive paneling

• Rabbet Plane (Br. Rebate Plane)

The word *rabbet* derives from the French: *rabattre*, meaning "to beat down" or "reduce," and as such is also the origin of the financial term *rebate*. The American spelling is closer to the original and preserves a useful distinction. Curiously, although spelled *rebate* in Britain, it is nonetheless pronounced *rabbet* everywhere.

A rabbet plane cuts a step-like section at the edge of a piece of wood. The salient feature of a rabbet plane is that its iron (blade) extends to the edge of the sole, enabling it to cut right into the corner and produce the rabbet. Wooden rabbet planes, which preceded the metal versions, are extremely common. Both forms are made with square and skewed irons, the skewed variety facilitating the clean cutting of cross-grain rabbets. Although the usual size is about 9 ½ in. long, there is great variety in length as well as width.

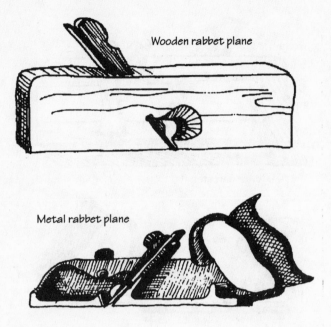

Wooden rabbet plane

Metal rabbet plane

• Rabbet Plane, Adjustable

The adjustable rabbet plane, also called a "cabinetmaker's rabbet plane," is made with sides that are perfectly square to its sole so that it will lie flat on either side and can be used as a right-hand or left-hand plane. The front part is adjustable, enabling the mouth to be made smaller or larger, or even be completely removed, allowing the plane to work right into a corner.

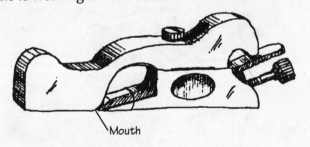

Mouth

• Rabbet Plane, Bull-nose

A bull-nose plane is so called because of the shortness of the section in front of its iron, which allows it to be used

very close to an obstruction. The bull-nose rabbet plane is designed to trim obstructed rabbets.

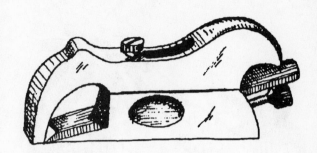

• Rabbet Plane, Carriagemakers Iron T-plane

The "T-plane" designation refers to the inverted T-shape of this tool. The narrowness of the stock allows it to be held in narrow spaces while still retaining a fairly wide cutting edge. T-planes, originally made of wood, are often characterized by relatively short soles, made in a variety of widths and curvatures—both lateral and longitudinal. Although originally adapted to the requirements of the carriagemaking trade, these planes are equally useful to the cabinetmaker.

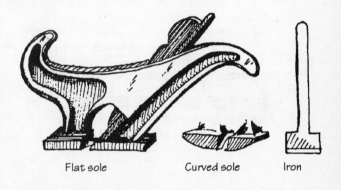

Flat sole   Curved sole   Iron

• Rabbet Plane, Carriagemaker's

The carriagemaker's rabbet plane is a long plane, similar in size and shape to a *jack plane*, but with an iron that

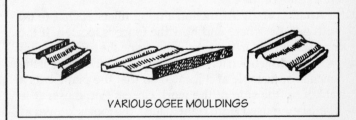

VARIOUS OGEE MOULDINGS

extends the width of the sole. This enables it to function as a bigger, handled, bench-style rabbet plane, useful for larger work than would be suitable for a regular rabbet plane.

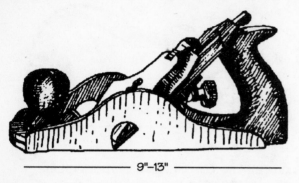

9"–13"

• **Rabbet Plane, Filletster** (*Br. Fillister Plane*)
Filletster is the recent Americanization of the older British term *fillister*, the origin of which is obscure. The filletster is another plane dedicated to cutting grooves across the surface of a board. The precise relationship of the various groove-cutting planes to the kinds of work they do is illustrated below.

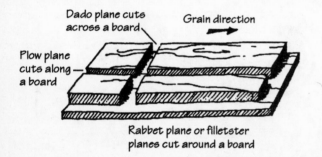

Dado plane cuts across a board

Grain direction

Plow plane cuts along a board

Rabbet plane or filletster planes cut around a board

The chief difference between a regular rabbet plane and a filletster is that the latter has a fence, sometimes integral (as in older, wooden models) and sometimes adjustable, removable, and capable of being used on either side of the tool.

An important distinction among filletsters has to do with exactly where and how the fence is located. If the fence is formed in the sole of the plane the tool is known as a "standing filletster" and its depth of cut is fixed. If the fence is adjustable and located so that the plane cuts a rabbet on the side of the work closest to the user (the plane being used from right to left), the tool is known as

a "moving filletster." If an adjustable fence is located so that the plane forms a rabbet on the side of the work farthest away from the user, it is known as a "sash filletster," since such tools are commonly used to cut the rabbet that houses the glass in window sash.

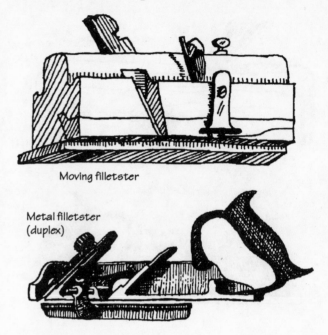

Moving filletster

Metal filletster (duplex)

• **Rabbet Plane, Side**
This little tool, made both in separate single-bladed left-hand and right-hand versions, as well as one that has two irons for cutting in either direction, is designed for trimming and cleaning up the sides or bottoms of rabbets, dadoes, and other kinds of grooves.

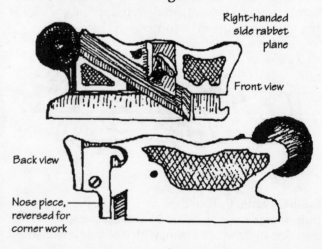

Right-handed side rabbet plane

Front view

Back view

Nose piece, reversed for corner work

• Router Plane (Old Woman's Tooth)

The router plane was designed to smooth the bottoms of grooves or other areas below the surface of the surrounding area. The early version of the tool, known as an old woman's tooth, was little more than a *chisel* fitted into a block of wood. In more modern metallic versions, the bit, which is made in different shapes for cutting across the grain and reaching into acute corners, is adjusted by a screw to the required depth, and the tool is guided freehand or with the help of fences that may be attached to its sole.

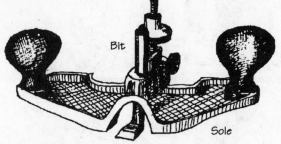

• Scraper Plane

The scraper plane is an elaboration of the basic *scraper*, being a scraper blade housed in a plane-like stock, and capable of being adjusted so that the angle at which it approaches the work can be easily controlled.

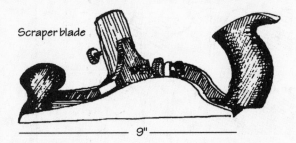

• Scrub Plane

The scrub plane has a wide mouth in the sole of its stock, through which projects a single, thick, and rounded iron (blade), ideally suited for the quick preparation of stock straight from the sawmill for planing with more finely adjusted planes, such as the *jack plane*.

• Spill Plane

Although woodworkers used spill planes, they were also a common household tool in the eighteenth and nineteenth centuries. Made in many forms, their function was to turn scrap lengths of wood into fire-lighting spills.

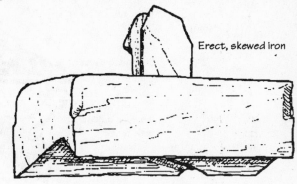

• Tonguing and Grooving Plane

This particular plane is an ingenious form of the older, paired *match planes*, combining in one tool the ability to cut both the tongue and the groove parts necessary for a tongue-and-groove joint.

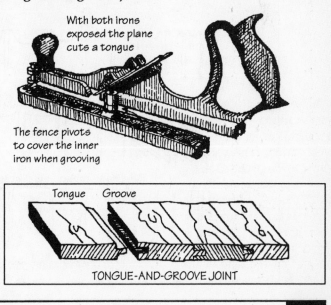

• Universal Plane

The universal plane is a multi-purpose plane, of which several models exist, designed to perform a whole range of planing jobs, thus making it unnecessary to carry around a large selection of individual special-purpose planes. Some of the tools subsumed by the universal plane include the *plow*, the *dado*, the *rabbet*, the *beading plane*, the *chamfer plane*, the *filletster*, as well as a variety of *moulding planes*. What makes this feasible is the large number of adjustments possible in the form of remov-able arms, fences, and depth gauges, as well as the large assortment of different irons, some of which are shown here.

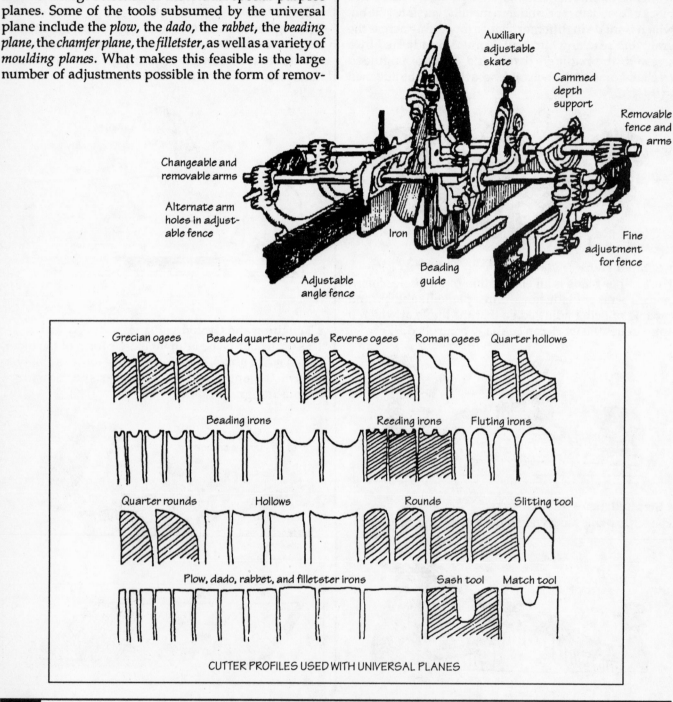

Auxiliary adjustable skate

Cammed depth support

Removable fence and arms

Changeable and removable arms

Alternate arm holes in adjustable fence

Iron

Fine adjustment for fence

Adjustable angle fence

Beading guide

CUTTER PROFILES USED WITH UNIVERSAL PLANES

Grecian ogees · Beaded quarter-rounds · Reverse ogees · Roman ogees · Quarter hollows

Beading irons · Reeding irons · Fluting irons

Quarter rounds · Hollows · Rounds · Slitting tool

Plow, dado, rabbet, and filletster irons · Sash tool · Match tool

4. Moulding Planes

Moulding planes are (generally) thin-bodied wooden planes with variously shaped soles and matching irons (blades) for forming many fanciful (or practical) prismatic sections known as "mouldings." As the use of mouldings increased in the eighteenth century, so did the number and variety of moulding planes, until, by the end of the nineteenth century, every carpenter was liable to carry large sets. The invention of the *universal plane* and the development of mass-produced, machine-made mouldings made the moulding plane largely obsolete.

Many moulding planes were made in graduated sets, and so a very large number of varieties exists, including many different combinations of profile elements. Hundreds of thousands of these tools were made. Most are known by the name of the profile they cut. The following are some of the more common generic types:

> Astragal Plane
> Beading Plane
> Crown Moulding Plane
> Cavetto Plane
> Cove (or Scotia) Plane
> Hollow Plane
> Nosing Plane
> Ogee Plane
> Ovolo Plane
> Round Plane
> Sash Moulding Plane
> Side Snipe
> Snipe's Bill (or Quirk Moulding Plane)
> Table Planes

• Astragal

An astragal is a raised bead. While astragal planes were made in sets to cut astragal mouldings of various sizes, the astragal itself frequently forms one of the elements in a more complex moulding cut by a larger and more complex plane, such as an ogee with a bevel and astragal.

• Beading Plane

Possibly constituting the largest family of moulding planes, the beading plane, whose function is to cut a simple, recessed bead (as distinguished from the astragal, which is a raised bead), is made in a wide range of sizes and different forms. Examples of beading planes include those designed to cut single beads, multiple beads, and beads in different locations (at the edge or in the center of a workpiece).

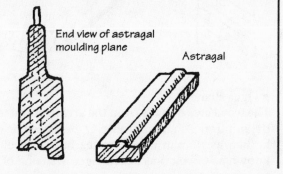

End view of astragal moulding plane

Astragal

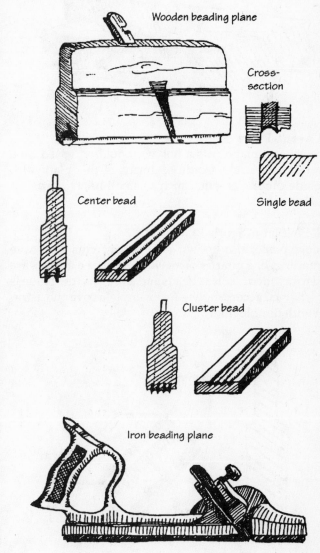

Wooden beading plane

Cross-section

Center bead

Single bead

Cluster bead

Iron beading plane

• Crown Moulding Plane:

A crown moulding is a large, complex moulding that was frequently made with the help of several small moulding planes, each forming one part of the whole profile. A crown moulding plane is the largest of all moulding planes, designed to cut the entire profile at once. The plane was frequently towed (by a rope attached to handles fixed in its side) and simultaneously pushed in the normal way.

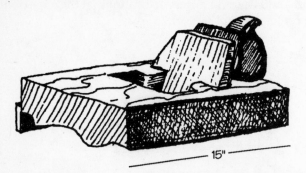

15"

• Cavetto Plane

The cavetto plane cuts a hollow moulding based on a quarter of a circle. Common forms of this plane also include other elements, such as small fillets (steps) or beads.

• Cove Plane (Scotia Plane)

A cove plane, also known as a "scotia," cuts a concave moulding, one quarter of an ellipse. Other elements are often included, such as beads and astragals (the plane is then named accordingly—for example, a cove and astragal with double fillet).

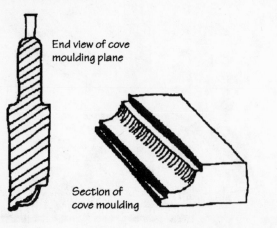

End view of cove moulding plane

Section of cove moulding

• Hollow Plane

The hollow plane, together with its partner the *round plane*, is rarely used on its own, but more often in conjuction with other tools to help create a profile for which there is no specific moulding plane to hand. Hollows and rounds were made in sets ranging from ⅛ in. to 2 in. in width. The curve of the hollow (or round) also varies (although 60° of arc is common), and it is therefore possible to create virtually any profile with just a few select hollows and rounds. Unlike all other moulding planes, which are named for the profile they cut, hollows and rounds are named for the shape of their sole—a hollow plane cutting a round profile, and a round plane cutting a hollow profile.

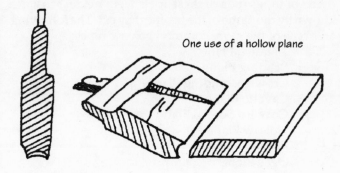

One use of a hollow plane

• Nosing Plane

The nosing plane cuts a round profile like that required at the front of stair treads.

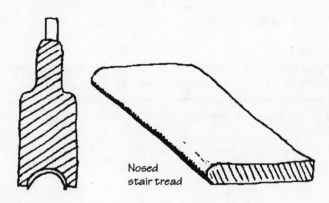

Nosed stair tread

• Ogee Plane

The word *ogee* derives from the architectural term *ogive*, denoting an S-shaped profile. Ogees based on circles are known as Roman ogees; ogees based on ellipses are known as Greek ogees. The ogee was one of the most common profiles in eighteenth- and nineteenth-century mouldings, and gave rise to dozens of varieties of ogee planes greatly varying in size and often combined with other elements.

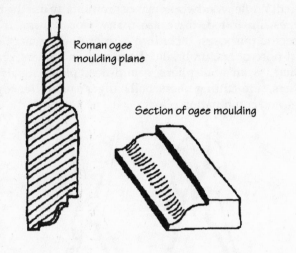

Roman ogee moulding plane

Section of ogee moulding

• Ovolo Plane
An ovolo plane forms a convex moulding based either on a circle (common ovolo) or an ellipse (Greek ovolo). It is one of the more common forms used for window sash.

• Round Plane
The round plane is the convex counterpart to the *hollow plane (q.v.)*.

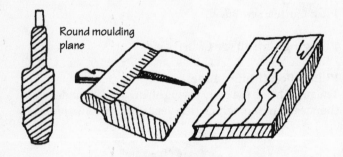

Round moulding plane

• Sash Moulding Plane
A sash moulding plane is designed to cut the moulded profile of wooden window sash. In America the sash moulding plane was commonly constructed so that it cut not only the moulded section but the rabbet in which the glass was set. In Britain two planes were required to make sash: the sash moulding plane and a separate *rabbet plane.*

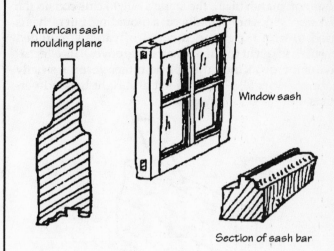

American sash moulding plane

Window sash

Section of sash bar

• Side Snipe
The side snipe is not strictly a moulding plane per se, being used only to trim the sides of quirks (thin vertical slits or grooves) that might form part of a moulding profile. In order to accommodate different grain and various possible locations of quirks, side snipes were made both left and right-handed.

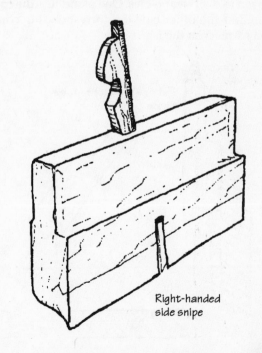

Right-handed side snipe

• Snipe's Bill (Quirk Moulding Plane)

Named for its similarity to the long bill of the snipe (a common marsh bird), the snipe's bill, like its cousin the *side snipe*, does not actually cut a moulding, but rather is used to trim the rounded edges of various moulded profiles adjacent to quirks (narrow grooves). Its alternative name, quirk moulding plane, is therefore not strictly accurate. It is also made in left and right-handed versions.

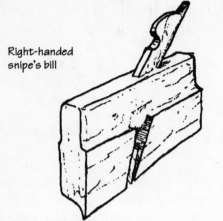

Right-handed snipe's bill

• Table Planes

Table planes are used in matched pairs to cut the rule joint used in drop-leaf tables. One plane cuts the convex portion and the other cuts the corresponding concave section formed on the leaf.

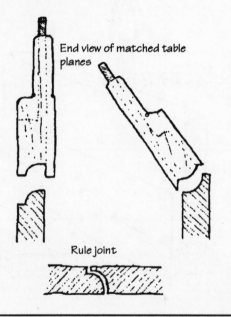

End view of matched table planes

Rule joint

Pliers

Pliers are a form of *pincers*, made with variously shaped jaws usually having parallel inner surfaces, designed for gripping securely whatever needs to be held or manipulated. While woodworkers most commonly use the two types illustrated, there are many more designed for specific purposes, including burner pliers, gas pliers, flat-nose pliers, umbrella pliers, weaver's pliers, chain pliers, swan's-bill pliers, round-nose pliers, insulated pliers, side-cutting pliers, bell hanger's pliers, fence pliers, and combination pliers.

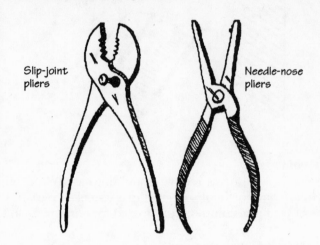

Slip-joint pliers

Needle-nose pliers

Plough Plane: see *Plane, Special-purpose.*

Plow Plane: see *Plane, Special-purpose.*

Plug Cutter: see *Bit.*

Plugging Chisel: see *Chisel, Cold.*

Plumb Bob (Plumb Line)

The plumb bob is a heavy weight, originally made of lead (hence the name, which derives from the Latin *plumbum*

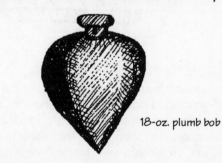

18-oz. plumb bob

meaning "lead"), suspended on a line. The tool may be referred to either as a plumb bob or a plumb line. Since the weight causes the line to hang vertically, this tool is used to mark a point immediately below another.

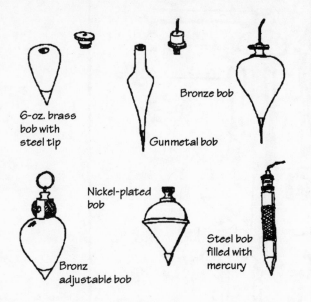

6-oz. brass bob with steel tip

Bronze bob

Gunmetal bob

Nickel-plated bob

Bronz adjustable bob

Steel bob filled with mercury

Plumber's Chisel: see *Chisel, Wood, Special-purpose.*

Plumber's Gouge: see *Chisel, Wood, Gouges.*

Plumb Level: see *Level.*

Plumb Rule
A plumb rule is a long piece of straight wood held vertically with a *plumb bob* suspended from the top end.

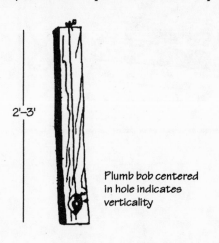

2'–3'

Plumb bob centered in hole indicates verticality

When the plumb bob is centered over a hole cut in the lower end, the tool may be used to measure the perpendicularity of anything it is held next to, such as a post or a wall. The invention of the *spirit level* rendered this tool obsolete.

Pocket Chisel: see *Chisel, Wood, Bench.*

Protractor
A protractor is an instrument, usually flat, semi-circular, and graduated in degrees, used for measuring angles. It frequently forms part of other tools such as *levels* and *bevels.*

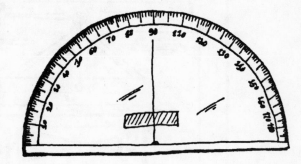

Pry Bar
A pry bar is a length of straight steel with a flat blade at each end, one end being offset, used for removing nails, separating pieces of wood, and general dismantling.

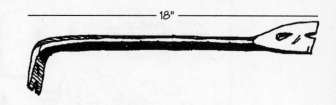

18"

Push Drill: see *Drill (1).*

Putty Knife: see *Knife.*

"THE FORE PLANE," based on an illustration by Moxon, 1703

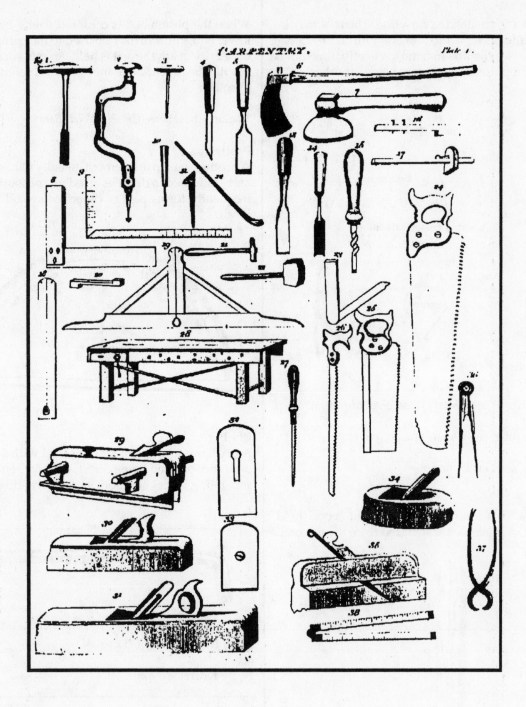

CARPENTRY.

Quirk Moulding Plane: see *Plane, Moulding.*

facing page:
Tools of the Carpenter and Joiner, from *The Circle of the Mechanical Arts,* by Thomas Martin, 1813

1. auger, 2. brace, 3. gimlet, 4. mortise chisel, 5. firmer chisel, 6. adze, 7. axe, 8. trysquare, 9. carpenter's square,
10. brad awl, 11. hook pin, 12. crow, 13. socket chisel, 14. gouge, 15. turnscrew, 16. mortise gauge, 17. marking gauge,
18. plumb rule, 19. plumb level, 20. bench hook, 21. hammer, 22. mallet, 23. bevel, 24. ripsaw, 25. tenon saw,
26. compass saw, 27. keyhole saw, 28. bench, 29. plow plane, 30. jack plane, 31. trying plane,
32. plane iron, 33. capiron, 34. smooth plane, 35. moulding plane,
36. dividers, 37. pincers, 38. folding rule.

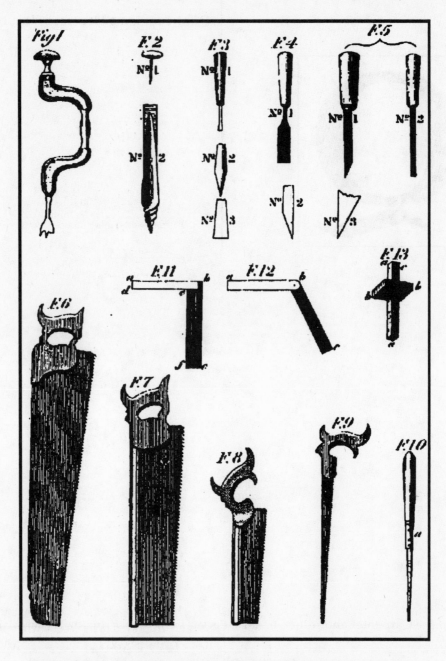

Tools of the Joiner, from *Mechanical Exercises,* by Peter Nicholson, 1812

1. brace, 2. gimlet, 3. brad awl, 4. firmer chisel, 5. mortise chisel, 6. handsaw,
7. tenon saw, 8. sash saw, 9. compass saw, 10. keyhole saw.

Rabbet Plane: see *Plane, Special-purpose.*

Rasp
A rasp may be defined as a very coarse *file* with raised points instead of lines for cutting. Like the file, rasps are made in different degrees of coarseness and various shapes. They are used for rapid removal of wood, especially in non-rectilinear work such as the shaping of curved legs or chair arms.

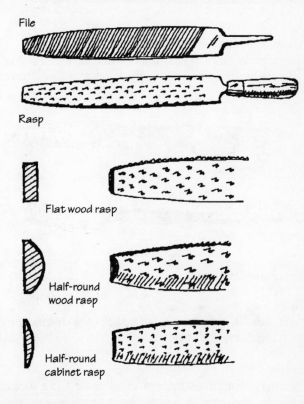

File

Rasp

Flat wood rasp

Half-round wood rasp

Half-round cabinet rasp

• Rasp, Horse
Horse rasps are the biggest and coarsest rasps. Although most other rasps, with the exception of the *shoemaker's rasp*, are intended to be used with handles fitted on tangs formed at one end of the rasp, the horse rasp is usually made square at both ends.

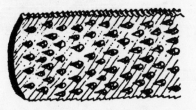

• Rasp, Lastmaker's (Shoemaker's or Shoe Rasp)
This common tool, sometimes sold as a "combination file," is part rasp and part *file*. Both sides of this handle-less tool are cut with file teeth and rasp teeth, one side being round and the other flat.

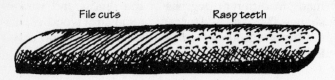

File cuts   Rasp teeth

• Rasp, Woodcarver's (Rasp Riffler)
A woodcarver's rasp is a *riffler* with a surface cut like a rasp—rifflers properly being, by definition, tools that are cut like a *file*.

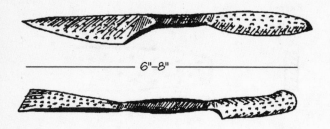

6"–8"

Razor Plow: see *Knife*.

Rebate
Rebate is the British spelling of *rabbet*, for a discussion of which see *Plane, Rabbet*.

Reamer
A reamer, sometimes known as a "rimer," is a tool for widening and truing holes. There are many reamers designed for use with metal. A wood reamer designed to be used in a *brace* is also referred to as a "bit stock taper reamer"—*bit stock* being an old term for a *brace*.

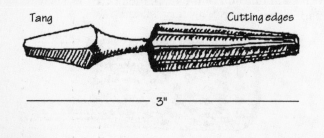

Tang   Cutting edges

3"

• Reamer, Wheelwright's
The wheelwright's reamer is much larger than the average wood reamer, since its function is to enlarge the hub holes of wooden wheels. The design remained essentially unchanged from the Middle Ages until the demise of wooden wagons early in this century.

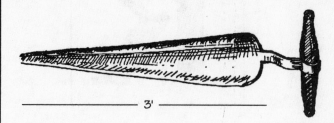

3'

Riffler
A *riffler* is a curved tool used by carvers and sculptors, usually finished like *file*. Woodcarvers often refer to their rifflers as a "riffler files" to distinguish them from rifflers cut like *rasps*, properly known as *woodcarver's rasps*—which is actually a contradiction in terms. (*See File, Woodcarver's*.)

Ripping Bar (Wrecking Bar)
A ripping, or wrecking, bar is a long, goose-necked steel bar, usually hexagonal in section, and toughened to withstand breaking. It is used by carpenters for all kinds of prying and levering.

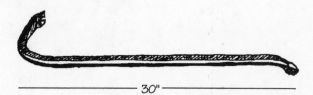

30"

Ripping Chisel: see *Chisel, Wood, Special-purpose*.

Ripsaw: see *Saw*.

Round Plane: see *Plane, Moulding*.

Round-nose Chisel: see *Chisel, Cold*.

Router Plane: see *Plane, Special-purpose*.

Rule
A rule is a graduated strip of wood or metal used for

linear measuring. The distinction between *rule* and *ruler* is that the former is principally used for measuring, while the latter more properly refers to a drawing instrument for guiding the pen or pencil in forming straight lines.

• Rule, Architect's

The architect's rule is commonly triangular in section, which gives it two additional faces on which a variety of scales and different calibrations are marked.

• Rule, Bench

A bench rule and its close relative the *yardstick* are the common rules used by woodworkers at the bench. They are usually graduated in ⅛-in. and ¹⁄₁₆-in. increments. The best rules are made of maple or hickory (woods little prone to warp) and are brass-tipped to protect their ends.

• Rule, Blacksmith's Hook-and-handle

The blacksmith's rule is made from rolled brass so that measurements can be taken on hot pieces of iron. The hook aids in keeping the rule against the work.

• Rule, Blindman's

The blindman's rule is a variety of the *folding rule*, being provided with extra-large figures and generally made a little wider for easier use.

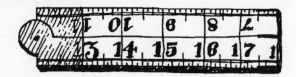

• Rule, Board

Board rules are used for computing the number of board feet in a given piece of wood. A board foot is the amount of wood comprised by a piece 1 ft. long by 1 ft. wide and 1 in. thick. If the thickness doubles, the width must halve for the measurement to remain constant. The rule takes into account the various widths and thicknesses that wood is commonly sawn into, and provides values for different lengths.

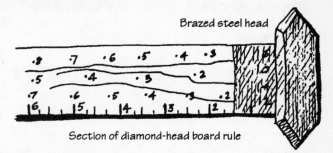

Section of diamond-head board rule

• Rule, Caliper

Caliper rules combine the function of *calipers* and short rules. Unlike most American rules, the graduations generally read from left to right.

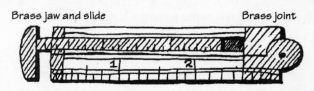

• Rule, Folding

The folding rule was once the most common rule used by carpenters. It was succeeded by the *zig-zag rule* and then by the metal *tape measure*. The basic folding rule is 2 ft. long when extended and folds four times. This is known as a "two-foot four-fold rule." There are also, however,

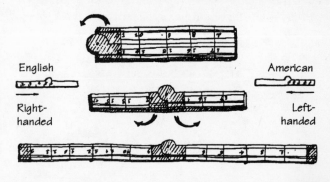

one-foot four-fold rules, two-foot two-fold rules, three-foot four-fold rules, and four-foot four-fold rules. American folding rules are graduated left-handedly; that is, the numbers begin at the right-hand end, whereas British folding rules are usually right-handed, with numbers beginning at the left end.

• Rule, Log
A log rule is used by lumberjacks and sawmill operators to compute the amount of wood in any given log. The rules are graduated according to various systems, one of the most common being Doyle's System.

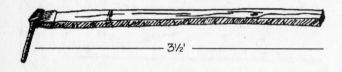

$3\frac{1}{2}'$

• Rule, Patternmaker's Shrinkage
Shrinkage rules are graduated with a variety of scales that reflect how much different metals shrink as they cool. In order for the patternmaker to produce a casting of a required size, he must make a mould somewhat larger; the shrinkage rule is therefore longer than a regular rule of the same nominal length.

• Rule, Ten-foot
A ten-foot rule is an old measuring device used by carpenters and builders. It consists of a 1-in. square rod, divided along its length into feet and inches. A five-foot version was also common.

• Rule, Yardstick
A wooden rule measuring 3 ft. long (a yard), frequently dispensed as a premium, is known as a yardstick.

• Rule, Zig-zag
The zig-zag rule consists of sections of wood, of various lengths, jointed together in such a way as to allow complete or partial extension or contraction. The most common (fully extended) lengths are 6 ft. and 8 ft., although 2-ft., 3-ft., 4-ft., and even 5-ft. rules have also been made.

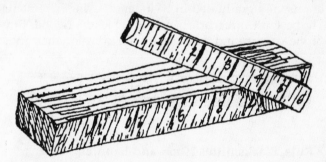

Gile's Drag Sawing Machine, from *Grimshaw on Saws*, by Robert Grimshaw, 1880

Carpenter Holding Framing Square and Dividers, the Frontispiece from
The Description and Use of the Carpenter's Rule, by John Brown, 1688

Sandpaper Holder

A sandpaper holder is any one of numerous devices for holding sandpaper securely during use. Simple holders may be mere blocks of wood; more elaborate devices are often made of other materials, fitted with a variety of clips and paper-securing mechanisms.

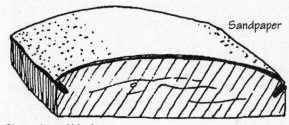

Sandpaper

Slotted wood block

Sash Filletster: see *Plane, Special-purpose, Rabbet.*

Sash Moulding Plane: see *Plane, Moulding.*

Saw

The word *saw* is very old; it derives from the pre-Teutonic root: *sek,* meaning to cut. Early saws were made of stone or bone, but now the word implies a blade, bar, or rod made of metal, usually with teeth formed along one edge, although there are some saws made of abrasive material that cuts without teeth. Saws used in woodworking include the following, described in alphabetical order:

> Back Saw (Br. Tenon Saw)
> Bow Saw (Cabinet Saw, Turning Saw, or Web Saw)
> Bracket Saw (Fretsaw or Scroll Saw)
> Buck Saw
> Compass Saw
> Coping Saw
> Crosscut Saw
> Dovetail Saw
> Gauge Saw
> Gent's Saw
> Hacksaw
> Handsaw
>> Crosscut Handsaw
>> Panel Saw
>> Ripsaw
> Inside-start Saw

The Illustrated Encyclopedia of Woodworking Handtools Instruments & Devices

Jeweler's Saw
Joiner Saw (Bench Saw)
Keyhole Saw
Miter-box Saw
Pit Saw
Plumber's Saw
Pruning Saw
Ship Carpenter's Saw
Veneer Saw

• Back Saw (Br. Tenon Saw)

The back saw is a medium-size bench saw, used for fairly exact joinery. Its name reflects the fact that it is fitted with a stiffening rib along its back—something that is necessary because, since its teeth are designed to cut on the push stroke, the thin blade would otherwise buckle. In Britain it is referred to as a tenon saw, since it is the saw best adapted for cutting the tongued parts (known as tenons) of mortise-and-tenon joints.

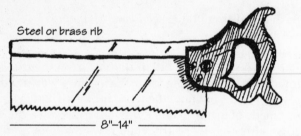

Steel or brass rib

8"–14"

Saw Teeth: One of the salient points about saws is how many teeth they have. The number and form of the teeth determine what kind of sawing the saw is best suited for. The actual quantity of teeth is usually expressed in terms of "points per inch," which is always one more than the actual number of teeth.

1"   1"

7 points per inch   6 teeth per inch

• Bow Saw (Cabinet Saw, Turning Saw, or Web Saw)

The term *bow saw* indicates a saw blade held in a bow-like frame, and includes a variety of saws used for different purposes. That most such saws have rotatable blades classifies them further as turning saws. Since they were once the predominant form of cabinetmaker's saw (and indeed still are in certain parts of the world), they were also known generally as cabinet saws. Different types, such as the chairmaker's bow saw, the felloemaker's bow saw, or the German bow saw, largely reflect various blade sizes.

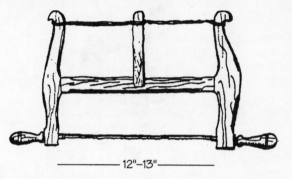

12"–13"

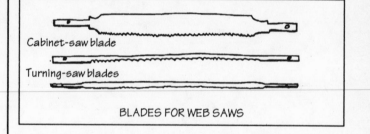

Cabinet-saw blade

Turning-saw blades

BLADES FOR WEB SAWS

• Bracket Saw (Fretsaw or Scroll Saw)

The three names by which this saw is known all describe the kind of work it is designed to do, namely, intricate curved cutting of relatively thin wood. This kind of work is known as fretwork, commonly used to ornament brackets, and is therefore also often known as scrolling or

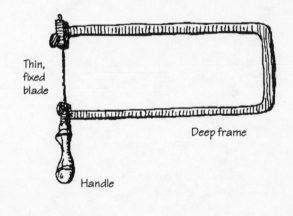

Thin, fixed blade

Deep frame

Handle

scrollwork. The saw's deep frame allows the thin blade to work far in many directions.

• Buck Saw

The buck saw is used for sawing firewood while supported in a saw buck—a type of *saw horse* with extended sides forming an X-shape. Ten-dollar bills are known as sawbucks because the Roman numeral for ten (*X*) with which early American banknotes were marked, resembled the shape of the saw buck.

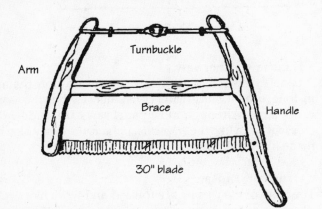

Arm
Turnbuckle
Brace
Handle
30" blade

• Compass Saw

The narrow blade of the compass saw allows it to cut curves (as described by a compass). Older compass saws were made with no set (the bending outward of the teeth to create a kerf, or cut, wide enough to allow the rest of the blade to pass through). Instead, the back of the blade was made thinner than the edge in which the teeth were cut. Modern compass saws are typically made with eight to ten points per inch, whereas older saws frequently only had five points per inch.

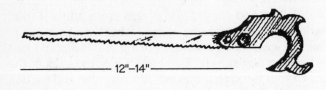

12"–14"

• Coping Saw

Coping describes the process of cutting the end of a piece of wood to match the curved profile of another piece to which it is joined. The coping saw therefore has a very thin blade to allow it to make tight turns. It is usually also possible to rotate the blade within its frame.

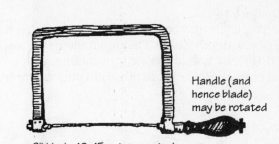

Handle (and hence blade) may be rotated

6" blade, 10–15 points per inch

• Crosscut Saw

Crosscut saws (not to be confused with *crosscut handsaws*) are large saws, often designed as two-man saws, intended for sawing whole trees. The one-man crosscut saw is typically about 3 ft. to 4 ft. long and made with plain teeth. It is used mainly for sawing felled trees into logs. The two-man crosscut saw may be as long as 7 ft. It has very large teeth, designed for felling trees, arranged in repeated groups of three: two "cut-off" teeth and one "clearer" tooth.

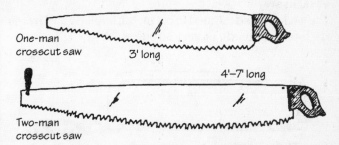

One-man crosscut saw
3' long
4'–7' long
Two-man crosscut saw

• Dovetail Saw

The dovetail saw is essentially a smaller version of the back saw made with much finer teeth suited to extremely fine work, such as the precise cutting of dovetail joints.

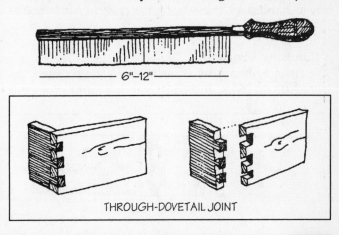

6"–12"

THROUGH-DOVETAIL JOINT

• Gauge Saw

The gauge saw may be any regular *handsaw* fitted with an adjustable depth gauge. The gauge allows the saw to be used for exact shouldering, dovetailing, or any other kind of work where the depth of cut must be controlled.

• Gent's Saw

Tools made somewhat smaller than the regular size were popular in the nineteenth century, being considered suitable for gentlemen hobbyists. A gent's saw usually implies a smaller *back saw*, but it is sometimes used (erroneously) to refer to a *dovetail saw*.

• Hacksaw

The hacksaw is designed to cut metal; its blade and frame are therefore made of very strong steel.

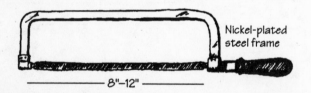

• Handsaw

Although all the saws in this book are used by hand, by convention only three types of saw are classified specifically as handsaws, these being the *crosscut handsaw*, the *ripsaw*, and the *panel saw*.

• Handsaw, Crosscut

The crosscut handsaw is made in a variety of lengths from 20 in. to 26 in., and with a varying number of teeth

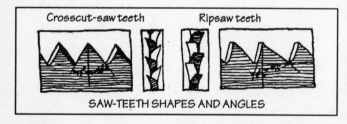

SAW-TEETH SHAPES AND ANGLES

ranging from eight points per inch to twelve points per inch (although formerly as many as fifteen points per inch was common). The particular way in which this saw's teeth are formed suit it especially for cutting across the grain. Teeth formed this way are, in fact, known as "crosscut teeth."

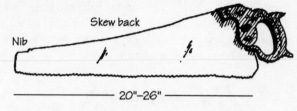

• Handsaw, Panel Saw

A panel saw is a small handsaw, usually no more than 20 in. long, used for its greater convenience when the work is not large. Although most panel saws are made with crosscut teeth like the *crosscut handsaw*, they may also be made like small *ripsaws*.

• Handsaw, Ripsaw

The teeth of the ripsaw are formed and work like small *chisels*. In comparison, the teeth of the *crosscut handsaw* are knife-like. Ripsaw teeth are designed to cut along, rather than across, the grain. For this reason they are generally fewer in number per inch.

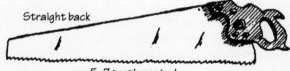

• Inside-start Saw

The toothed and curved end of this saw enables it to be started in the middle of a board rather than at one end. This is useful when sawing flooring that is already laid and where there are no free ends. Once the back of the saw has cut through the wood, it is reversed and used in the usual manner.

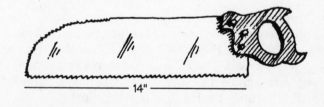

• Jeweler's Saw

The jeweler's saw is a fine saw similar in design to a *fretsaw*, but generally much smaller.

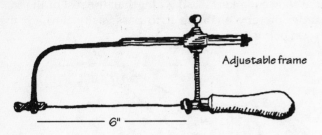

Adjustable frame

6"

• Joiner Saw (Bench Saw)

The joiner saw is a cross between a spineless *back saw* and a *handsaw*. Usually about 17 in. long and made of spring steel, it is useful for general joinery—a class of work finer than carpentry but not as exacting as cabinetmaking.

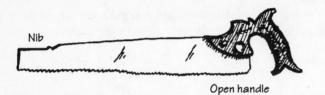

Nib

Open handle

• Keyhole Saw

The narrow-bladed keyhole saw used to be called a "turning saw" but should not be confused with the *turning saw* described under *web saw* above. It is usually provided with several interchangeable blades, each of which enables it to be used for different classes of curved work—cutting keyholes and similarly restricted sections being typical. A saw provided with interchangeable blades is said to be "nested."

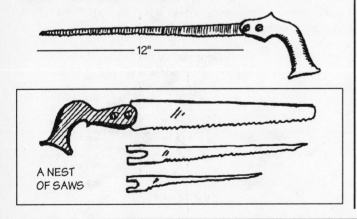

12"

A NEST
OF SAWS

• Miter-box Saw

The miter-box saw is an enlarged *back saw* intended to be used with a *miter box*. Often sold as part of a *metal miter box*, the spine enables the saw to run in special guides.

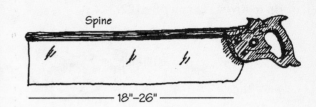

Spine

18"–26"

• Pit Saw

The pit saw was the saw used by sawyers working in pits before the days of sawmills. Logs were placed over large pits, while a top-sawyer (standing above the pit) held the tiller and a bottom-sawyer, working at the bottom of the pit, held the lower end to produce planks from the log.

Tiller

7' blade

• Plumber's Saw

The plumber's saw is a double-edged saw—one side for cutting pipe and the other for cutting wood.

Metal-cutting teeth

Wood-cutting teeth

• Pruning Saw

The pruning saw is a double-edged saw intended for gardening and tree work.

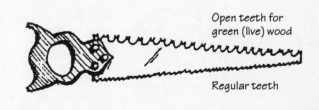

Open teeth for
green (live) wood

Regular teeth

• Ship Carpenter's Saw

The ship carpenter's saw is a form of *handsaw* heavier than a regular handsaw. In contrast to the skew backs that are a feature of some *crosscut handsaws*, it is always made with a straight back . It may have either crosscut teeth or rip teeth (*see box on page 122*).

• Veneer Saw

A veneer saw is a very small saw, only about 4 in. to 5 in. long, used for cutting thin sheets of veneer. The edge of the saw being curved, it is possible to start the cut in the center of a piece of veneer, which might crumble if sawn from one end.

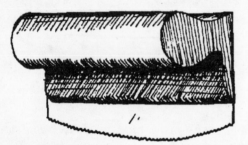

Saw Horse

The saw horse is standard equipment in most woodworking shops, as well as on many construction sites. Often used in pairs, the saw horse supports long boards being sawn or otherwise worked on.

Saw Jointer

The saw jointer is a device for holding a small *flat file* that is run along the teeth of a saw to file them all to the same height—this process being known as "jointing."

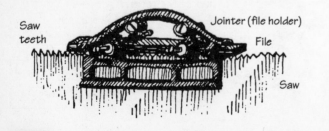

Saw teeth   Jointer (file holder)

File

Saw

Saw Set

A saw set is a tool, much like *pliers*, that is placed over the teeth of a saw to bend them slightly outward in order so that they cut a kerf (slot) wider than the rest of the saw blade, thereby enabling it to pass easily through the wood being sawn.

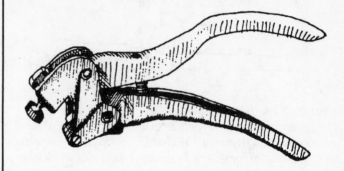

Saw Wrest

A saw wrest was the tool used to bend a saw's teeth to the required set before the invention of the *saw set*.

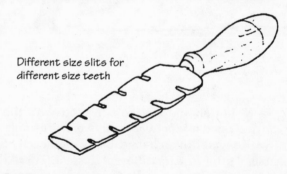

Different size slits for different size teeth

Scorp

The scorp, or "scoop," is a cross between an *adze* and a *drawknife*. It is used for hollowing out depressions such as concave chair seats or bowls. It is usually two-handled. (*See also Inshave.*)

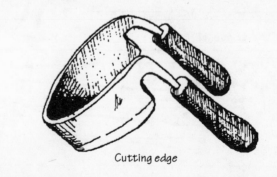

Cutting edge

Scotia: see *Plane, Moulding.*

Scraper

A scraper is used to scrape away the ridges sometimes left after planing, and may also be used to smooth surfaces difficult to plane by virtue of the nature of the grain, or because the shape cannot be reached by a *plane.*

• Scraper, Adjustable

An adjustable scraper is a plain *cabinet scraper* with an attached but adjustable handle. The handle makes it possible to use more of the scraper than would be possible if held by hand, since the scraper becomes smaller after every sharpening.

Adjusting screw

Handle

Scraper blade

• Scraper, Box

A scraper box is an especially stout scraper used for rough cleaning, as opposed to the fine scraping performed by a *cabinet scraper.*

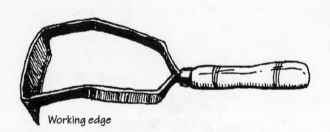

Working edge

• Scraper, Butcher Block

The butcher block scraper is a very heavy-duty scraper used on hardwood surfaces such as butcher block—material composed of many pieces of hardwood arranged so that the longer wearing end-grain forms the uppermost surface.

• Scraper, Cabinet

Although apparently a very simple tool, consisting of a mere rectangle of high-quality steel, the cabinet scraper requires considerable skill to be sharpened and used properly.

All four edges may be sharpened for use

• Scraper, Cabinetmaker's

The cabinetmaker's scraper is a scraper blade set in a *plane*-like stock susceptible to various fine adjustments.

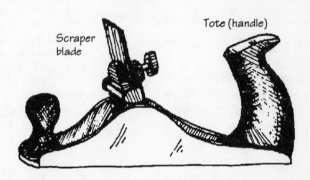

Scraper blade

Tote (handle)

• Scraper, Deck

The deck scraper is one of the largest scrapers, its blade being 5 in. across and its handle almost 2 ft. long.

• Scraper, Machinist's

The machinist's scraper is made in a variety of styles, all of which look similar to *files*. Their surfaces, however, are smooth; it is the edges that are designed to cut. The blades invariably taper to form a tang on which a handle is intended to be fitted.

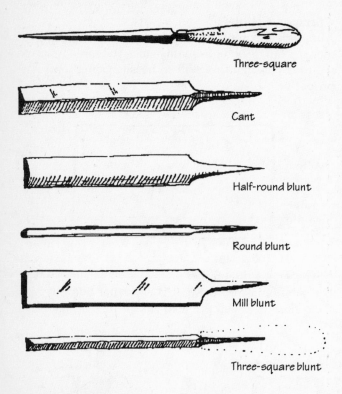

Three-square

Cant

Half-round blunt

Round blunt

Mill blunt

Three-square blunt

• Scraper, Moulding

Moulding scrapers are designed to smooth various moulding profiles, and are consequently made in a variety of matching shapes.

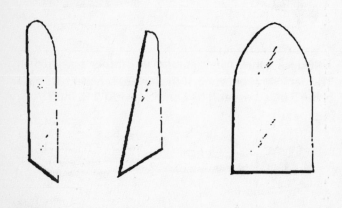

• Scraper, Oval

An oval-shaped *cabinet scraper*.

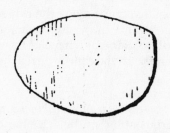

• Scraper, Swan's Neck

A *cabinet scraper* of particular shape used for smoothing curved surfaces. The constantly changing profile of this scraper allows it to be used on a variety of differently curved areas.

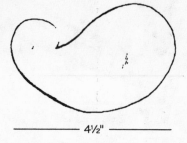

4½"

• Scraper, Veneer

Since veneer is usually very thin, a scraper intended to cut veneer must be capable of fine adjustment and be very carefully controlled. The veneer scraper is very similar to the *cabinet maker's scraper*, although considerably smaller.

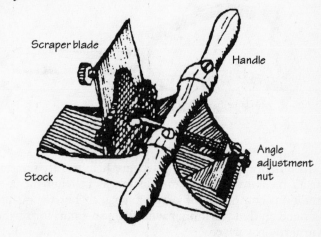

Scraper blade

Handle

Stock

Angle adjustment nut

• Scraper, Wall

The wall scraper is not used for wood; it is intended for cleaning old wallpaper off walls to be refinished.

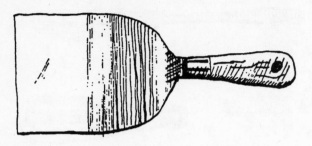

Scraper Burnisher

The process of turning the filed edge of the scraper so that it forms a tiny cutting edge is called "burnishing." A scraper burnisher consists of a hardened steel blade of various cross-sections, including round and triangular, designed to be rubbed with great pressure along the scraper's edge.

Scraper Plane: see *Plane, Special-purpose.*

Screw Box

A screw box is a tool for cutting a thread on a round piece of wood. Its counterpart, a *tap*, cuts a corresponding thread in a piece of wood so that the two pieces may be screwed together.

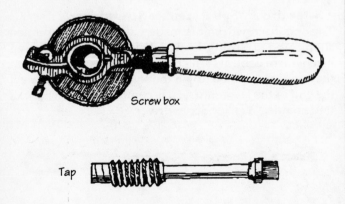

Screw box

Tap

Screwdriver

The screwdriver, unlike many other tools, is relatively new, having been common on woodworking benches only since the early nineteenth century. It was originally known as a "turnscrew." Aside from many different types, each designed for a particular trade or purpose, the design of screw heads also varies, necessitating different forms of screwdriver blades.

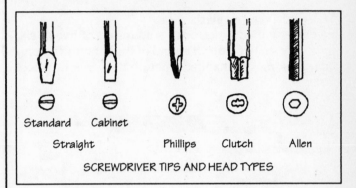

Standard Cabinet

Straight Phillips Clutch Allen

SCREWDRIVER TIPS AND HEAD TYPES

• Screwdriver, Brace

Any *brace* fitted with a *screwdriver bit* may be termed a brace screwdriver, but there are braces fitted permanently with a screwdriver for this single purpose.

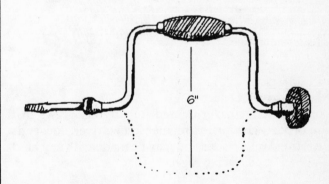

6"

• Screwdriver, Clock

The clock screwdriver is a small cabinet screwdriver, whose blade is no wider than its shank. This enables the screwdriver to turn screws located at the bottom of holes.

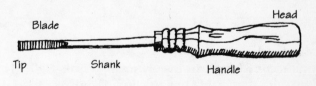

Blade

Head

Tip Shank Handle

• Screwdriver, Gunsmith's

The gunsmith's screwdriver is a short but stout screwdriver.

— 1" —

• Screwdriver, Jeweler's

Jeweler's screwdrivers, often sold in sets of five or six, have very small blades. The end of the handle is revolvable, making one-handed operation of this tiny tool easier.

Revolving head

Diameter of blade: .040"–.100"

Watch screwdriver

• Screwdriver, Locksmith's

The locksmith's screwdriver is short but very strong.

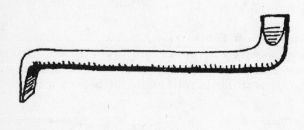

• Screwdriver, Offset

An offset screwdriver is used for driving screws that cannot be reached with a regular screwdriver. Both ends are formed into blades and may be made with any kind of tip (*see box on facing page*).

• Screwdriver, Pocket

Pocket screwdrivers are merely small, pocket-size tools, often with retractable tips or clips. They are a typical example of the apparently endless ingenuity that has been lavished on even the simplest of tools.

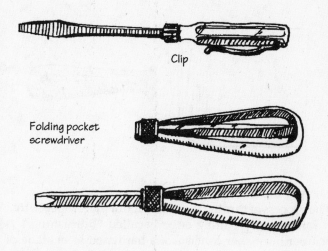

Clip

Folding pocket screwdriver

• Screwdriver, Ratchet

The ratchet screwdriver is a more convenient form of *offset screwdriver*. Equipped with a ratcheting device, and often with interchangeable tips, it allows screws located in tight places to be driven in or out without constantly having to reposition the screwdriver.

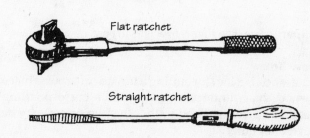

Flat ratchet

Straight ratchet

• Screwdriver, Spiral (Yankee Screwdriver)

The spiral screwdriver is made with a doubly spiraled shank and a spring-loaded handle. The ratcheting mechanism controls which way the tip is turned when the handle is pushed down the shank. Invented toward the end of the last century, it is commonly known by its trade name: "Yankee Screwdriver."

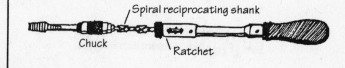

Spiral reciprocating shank

Chuck

Ratchet

• **Screwdriver, Undertaker's**
A short, stout screwdriver developed for the coffin-making trade.

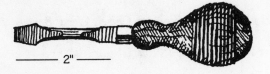
2"

Screwdriver Bit
A screwdriver bit, designed to work in a *brace*, is made in various patterns to fit different types of screw heads. The two kinds illustrated represent the most and least common varieties.

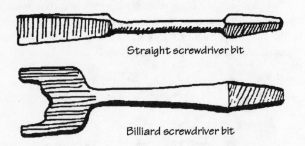
Straight screwdriver bit

Billiard screwdriver bit

Scriber
Although the term "scriber" generally has to do with writing, in carpentry it is used to mean a tool that marks by scratching.

• **Scriber, Marking Awl**
The marking awl scriber is a form of *awl* made entirely from steel tubing, with a knurled and nickle-plate shank into which the reversible point may be telescoped for safety when not being used.

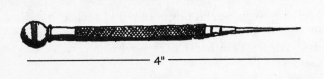
4"

• **Scriber, Timber (Timber Scribe)**
The timber scribe is a very old tool, having been used for centuries by those needing to identify by scoring all kinds of timber from raw logs to hewn beams. The kind

shown has an additional blade that cuts a shallow groove when pulled towards the user.

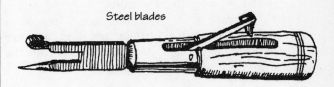
Steel blades

Scroll Saw: see *Saw.*

Scrub Plane: see *Plane, Special-purpose.*

Sharpening Stones
Many sharpening stones are still made from natural stone, but many others are so-called artificial stones, being man-made from various abrasive materials. Examples of the former include Arkansas and Washita stones (from America) and Turkey stones (from Europe). Examples of the latter include carborundum stones (made from silicone carbide), emery, and corundum (made from aluminum oxide).

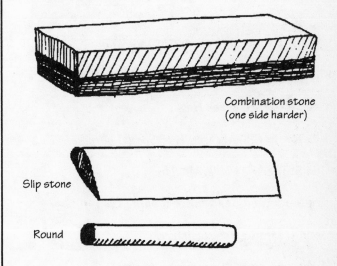
Combination stone
(one side harder)

Slip stone

Round

Until the nineteenth century, most stones were used with water as a lubricant. Thereafter, it became common to use oil, and as a result *oilstone* became synonymous with *sharpening stone*. The recent reintroduction of stones (notably from Japan), intended to be used with water as a lubricant, has necessitated the use of the term *water stone* to distinguish these stones from the so-called *oilstones*. Water stones may also be man-made or consist of natural stone quarried from specific locations.

Some stones are made in various shapes to accommodate differently shaped cutting edges, and most are made in different degrees of hardness. The harder the stone, the finer the edge it will impart.

Shavehook

A shavehook is a plumber's *scraper* used for shaping lead pipes.

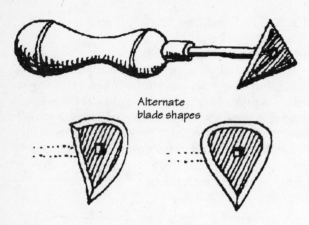

Alternate blade shapes

Shooting Block: see *Clamp, Miter-planing.*

Shooting Board: see *Chute Board.*

Side Hook: see *Bench Hook.*

Side Rabbet Plane: see *Plane, Special-purpose.*

Side Snipe: see *Plane, Moulding.*

Single-handed Beader: see *Beading Tool.*

Sledge Hammer: see *Hammer.*

Slick: see *Chisel, Wood, Special-purpose.*

Slitting Gauge: see *Gauge.*

Smooth Plane: see *Plane, Bench.*

Snipe's Bill : see *Plane, Moulding.*

Socket Firmer Chisel: see *Chisel, Wood, Bench.*

Socket Firmer Gouge: see *Chisel, Wood, Gouges.*

Socket Paring Chisel: see *Chisel, Wood, Bench.*

Spar Plane: see *Plane, Special-purpose.*

Spill Plane: see *Plane, Special-purpose.*

Spirit Level: see *Level.*

Splitting Gauge: see *Gauge.*

Spoke Pointer

A spoke pointer is a large *dowel pointer* used in a *brace* for forming a point on pieces of wood up to 2 in. in diameter. An old carpenter's saying celebrates this once rare tool: "A pointer loaned ne'er comes home."

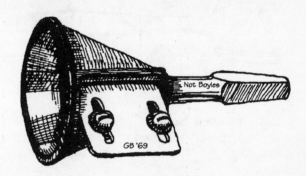

Spokeshave

The spokeshave may be considered a narrow *drawknife* set in a stock so that the depth of cut may be regulated. In fact, it is more like a small *plane*, capable of traveling over curved surfaces. One of its principal uses is the shaving and shaping of round pieces of wood such as handles—or spokes. Unlike a plane, it is usually operated by being drawn toward, rather than being pushed away from, the user. Until the Civil War, spokeshaves were made of wood.

• Spokeshave, Carriagemaker's Panel (Rabbet Spokeshave)

The iron (blade) of the carriagemaker's spokeshave extends to the corners of the sole and is sharpened on three sides so that it may be used to cut a step-like shave known as a rabbet.

• Spokeshave, Carriagemaker's
The carriagemaker's spokeshave is characterized by having straight and longer than usual handles.

• Spokeshave, Chairmaker's
In order to facilitate the concave shaping of chair seats, the chairmaker's spokeshave is made in a variety of curved forms.

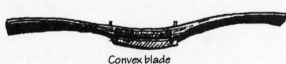

Convex blade

• Spokeshave, Chamfer
The chamfer spokeshave is a round-soled spokeshave fitted with adjustable "wings" that serve as 45° fences enabling the spokeshave to cut chamfers.

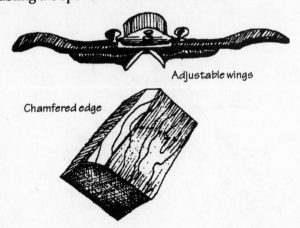

Adjustable wings

Chamfered edge

• Spokeshave, Iron
The iron spokeshave, made with a rounded or flat sole, is the common tool used by the majority of contemporary woodworkers. Its chief advantage over the older wooden variety is the ease with which the iron (blade) may be adjusted—although early iron spokeshaves lacked this feature.

Adjustment screws

• Spokeshave, Patternmaker's
The patternmaker's spokeshave exemplifies the earlier wooden form of the basic spokeshave.

Apple or boxwood handle

• Spokeshave, Razor-edge
This spokeshave gets its name from the fact that its iron (blade) is hollow-ground like the cutting edge of an old-fashioned straight razor. Since the front of the sole is adjustable, the thickness of the shaving may be easily controlled.

Adjustable mouth

• Spokeshave, Universal
The universal spokeshave's main feature is that its handles may be fitted in the sides and in the top of the stock. The

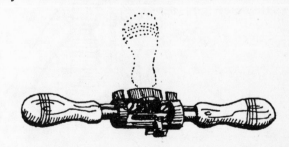

configurations thus made possible make this tool especially useful when working in restricted spaces.

Spud
A spud is a large tool used for debarking trees. It is similar to the *peeling chisel*, and was made in a variety of shapes and sizes.

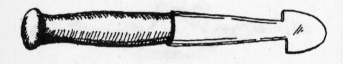

Square
The square is one of the oldest measuring tools. It is used for testing the squareness of sides, edges or surfaces. The basic square consists of a tongue or blade set at right angles in a thicker stock.

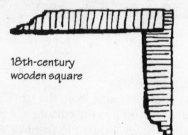

18th-century
wooden square

• Square, Adjustable Miter (Br. Adjustable Mitre Square)
The stock of the adjustable miter square is provided with an additional tongue that may be used as a *bevel*. Furthermore, the ends of the stock are finished at an angle of 45° to facilitate the laying out of miter joints.

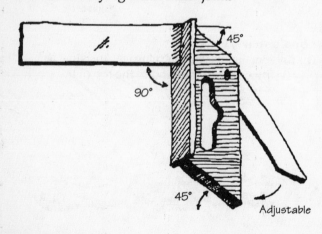

45°

90°

45°

Adjustable

• Square, Carpenter's (Framing Square)
The carpenter's square, known equally often as the framing square, is a large metal square marked with a variety of scales and and gradations designed to assist in the measuring and laying out of the various framing members that are used in house construction. It is especially useful in the construction of rafters and stairways and wherever differently angled cuts are required.

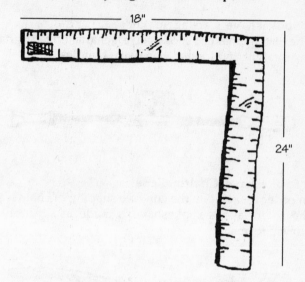

18"

24"

• Square, Combination
In America the term *combination square* refers to a steel tool much used by carpenters for measuring and marking both 90° angles and 45° angles. (*See also Try and Miter Square.*)

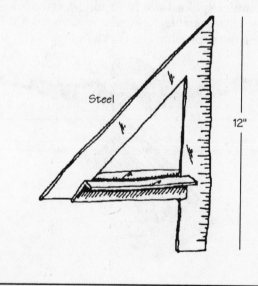

Steel

12"

• Square, Miter (Br. Mitre Square)

The miter square should properly be called a "try-miter" since it cannot measure or mark anything square, but rather is designed expressly for working with miters—which by definition are anything but square.

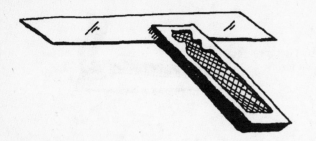

• Square, Trysquare

The trysquare is a simple tool made in a large range of sizes for trying (testing) the squareness of a board. Early trysquares were user-made entirely of wood. Nineteenth-century examples often had stocks made of rosewood, lined with brass. Contemporary tools are frequently all metal.

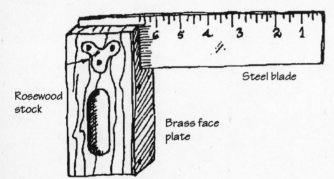

Rosewood stock

Steel blade

Brass face plate

• Square, Try and Miter (Br. Combination Square)

The stock of this square may be fixed at any point along its graduated tongue. It is so formed that 45° angles as

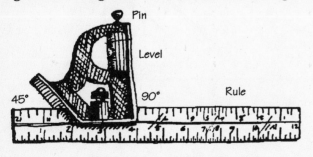

Pin

Level

45°   90°   Rule

well as 90° angles may be measured and marked—for which purpose it contains a removable steel pin that may be used as an *scriber*. Sometimes provided additionally with a centering attachment, this is a true combination tool, functioning as a *square, miter gauge, depth gauge,* and center finder.

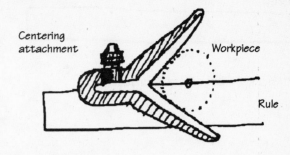

Centering attachment

Workpiece

Rule

Stair-gauge Fixtures

A pair of stair-gauge fixtures clamped to a steel *framing square* turns it into a device ideally suited for laying out stringers (the supporting members of staircases).

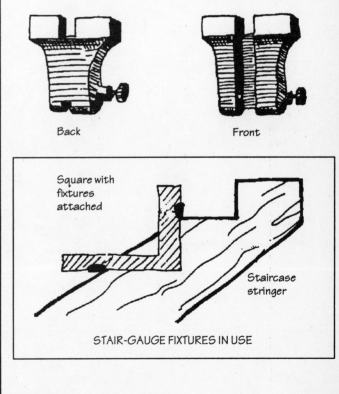

Back   Front

Square with fixtures attached

Staircase stringer

STAIR-GAUGE FIXTURES IN USE

Star Drill: see *Chisel, Cold.*

Straightedge

A straightedge is a perfectly straight length of steel, which may or may not be graduated, whose main function is to test for straightness or serve as a guide for another tool.

Steel

6"–72"

Sun Plane: see *Plane, Special-purpose.*

Surform Tool

Surform is a trade name for a family of tools designed for rapid wood removal. Made in various shapes, a surform tool consists of a steel blade pierced with many sharp-edged holes that work like a hollow *rasp*. The blade is held in a handled and lightweight aluminum body.

Plane-type surform

Table Planes: see *Plane, Moulding*.

Tack Claw

A tack claw is similar to a *nail claw* but much smaller, being intended only for tacks and small nails. The claw is driven under the head of the tack, which is then levered out.

Claw

Tape Measure

A tape measure was originally, as the name implies, a length of cloth tape for measuring linear distance. Instead of a length of marked cloth, wound by hand back into a leather case, tape measures now normally consist of spring-loaded metal tapes in lightweight metal or plastic cases. Tapes are made in a variety of sizes, ranging from lightweight tapes a few feet long to heavy duty tapes as long as 100 ft.

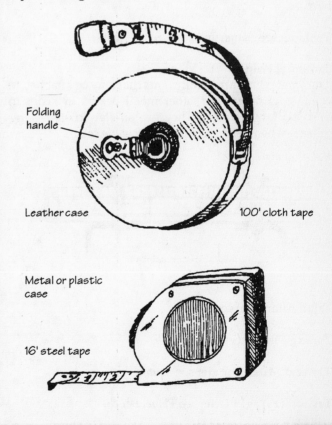

Folding handle

Leather case   100' cloth tape

Metal or plastic case

16' steel tape

Ten-foot Rod: see *Rule.*

Tenon Saw: see *Saw.*

Timber Scribe: see *Scriber.*

Tin Snips
Tin snips are included as representative of the large selection of metal-cutting tools, many of them in scissor form, frequently used by the woodworker.

Tonguing and Grooving Plane: see *Plane, Special-purpose.*

T-plane: see *Plane, Special-purpose, Rabbet.*

Trammel Points
Trammel points are metal points that may be attached to *rules*, steel *squares*, or other metal beams in order to describe large circles. They are similar to *stair-gauge fixtures* when used on a steel *carpenter's square.*

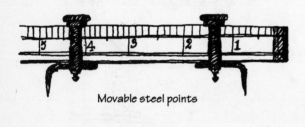

Movable steel points

Triangular File: see *File.*

Truing Plane: see *Plane, Bench.*

Try and Miter Square: see *Square.*

Try (or Trying) Plane: see *Plane, Bench.*

Trysquare: see *Square.*

Turning Chisel: see *Chisel, Wood, Woodturning.*

Turning Gouge: see *Chisel, Wood, Woodturning.*

Turning Saw: see *Saw.*

Tweezers:
Tweezers are a small gripping tool, related by function to *pliers* and *pincers*. They are used, in various sizes and patterns, by a wide variety of professions and trades from surgeons to pianomakers.

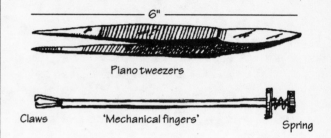

Piano tweezers

Claws   'Mechanical fingers'   Spring

Twibill (Br. Twybill or Two-bill)
A twibill is a small, two-headed axe used for cutting mortises *(see box on page 6).* The old pronunciation of *two* sounded the *w; bill* means a "curved cutting edge," as in the common agricultural pruning tool, the bill-hook.

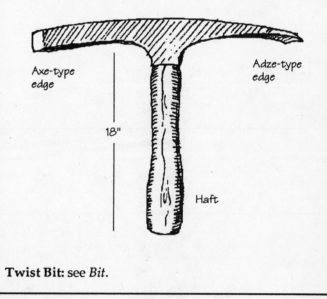

Axe-type edge   Adze-type edge

18"   Haft

Twist Bit: see *Bit.*

U

Undertaker's Screwdriver: see *Screwdriver*.

Universal Bevel: see *Bevel*.

Universal Bevel Protractor: see *Bevel*.

Universal Hand Beader: see *Beading Tool*.

Universal Plane: see *Plane, Special-purpose*.

Universal Spokeshave: see *Spokeshave*.

Upholsterer's Awl: see *Awl*.

Upholsterer's Hammer: see *Hammer*.

Utility Knife: see *Knife*.

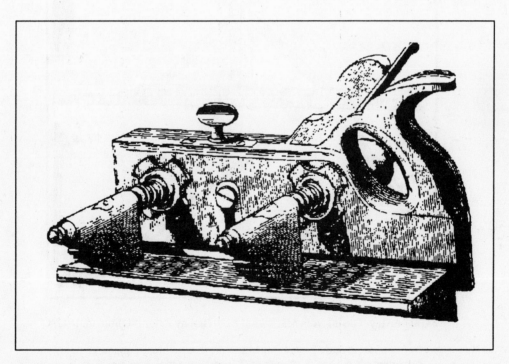

Plow Plane No 9B, Screwed Stems and Side Screw, from *The Illustrated Price List of Woodworking Tools*, of Alexander Mathieson & Sons, Ltd., 1899

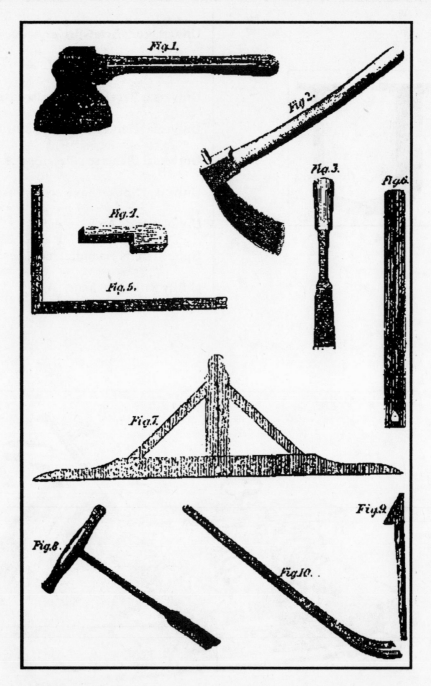

Carpentry Tools, from *Mechanical Exercises,* by Peter Nicholson, 1812

1. axe, 2. adze, 3. socket chisel, 4. mortice and tenon gauge,
5. carpenter's square, 6. plumb rule, 7. level, 8. auger, 9. hook pin, 10. crow.

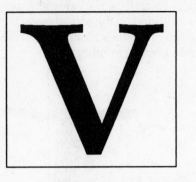

Veiner: see *Chisel, Woodcarving, Straight Gouges.*

Veneer Punch
A veneer punch is a large metal punch used to cut out small pieces of veneer to be used for repairing damaged sections of veneered work. The shape of the punch is irregular in order that the grain of the piece cut out may blend better when inserted.

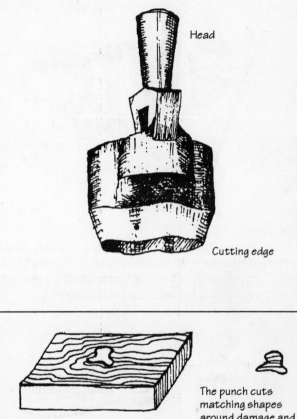

Head

Cutting edge

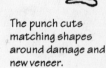

The punch cuts matching shapes around damage and new veneer.

VENEER PUNCH USE

Vise (Br. Vice)
A vise is a device consisting of two jaws that may be opened and closed by a lever, a cam, or more usually a screw. Vises used by woodworkers generally have wooden or wood-lined jaws in order to avoid marring the work. The word *vise* derives from the French: *vis*, meaning "screw"—by means of which early vises were operated. The American spelling is thus more accurate

and preserves a useful distinction between the tool and moral depravity.

• Vise, Coachmaker's

The coachmaker's vise, although having wood-faced jaws, is typical of the normal metal-working vise in having jaws that are higher than the surface of the bench on which it is mounted.

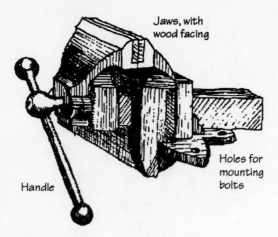

Jaws, with wood facing

Handle

Holes for mounting bolts

• Vise, Hand

There are many patterns of hand vises, which are small tools designed to be held in one hand for securely holding small objects such as bolts or screws. Many are made with tiny anvils (flat surfaces for pounding on).

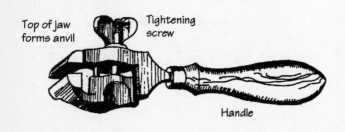

Top of jaw forms anvil

Tightening screw

Handle

• Vise, Pin

The pin vise is the smallest vise of all. It is provided with chuck-like jaws capable of securely holding small diameter objects, such as wire, tiny nails, or pins.

Steel chuck

Knurled handle

• Vise, Saw

A saw vise has especially long jaws designed to grip a *saw* just below its teeth while being sharpened. The saw vise may be mounted on the wall, it may clamp or be clamped to the edge of the bench or other surface, or it may even be held in turn by another vise.

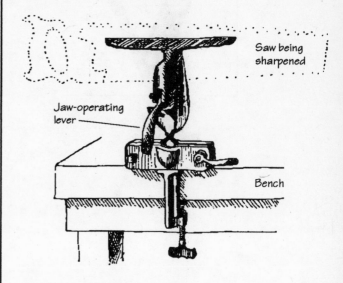

Saw being sharpened

Jaw-operating lever

Bench

• Vise, Woodcarver's

A woodcarver's vise is similar to the old form of carpenter's vise, consisting of a long vertical jaw that closes against the bench by means of a large wooden screw.

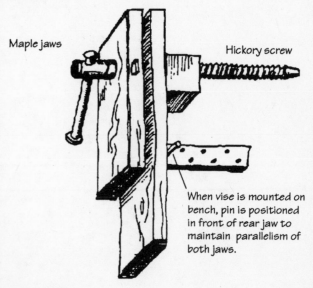

Maple jaws

Hickory screw

When vise is mounted on bench, pin is positioned in front of rear jaw to maintain parallelism of both jaws.

• Vise, Woodworker's

The woodworker's vise is made in a variety of patterns but is always fixed to, or is integral with, the workbench, and is invariably provided with wooden jaws whose top surfaces lie in the same plane as the benchtop. A stop is often provided in the front jaw.

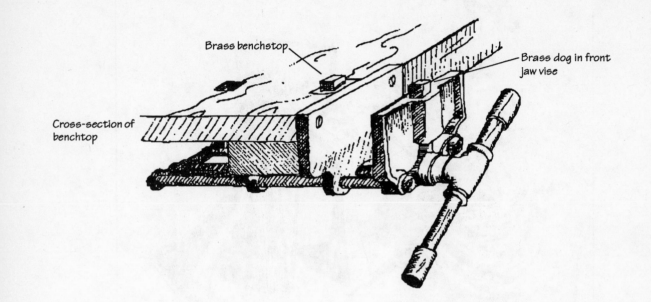

Brass benchstop

Brass dog in front jaw vise

Cross-section of benchtop

The Marquetry Sawyer, from *L'Art du Menuisier Ebéniste* (The Art of the Cabinetmaker), by Jacob-André Roubo, 1772

Wagonmaker's Drawknife: see *Drawknife*.

Wall Scraper: see *Scraper*.

Web Clamp: see *Clamp*.

Web Saw: see *Saw*.

Wedge
Wedges are used for splitting logs into smaller pieces. They are usually made of iron with steel tips, and may be sold by the pound according to weight. In Colonial America, oak wedges ringed with iron and called "gluts" were the common form.

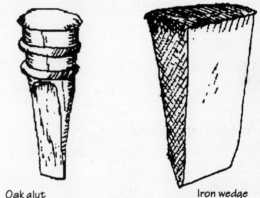

Oak glut   Iron wedge

Wheelwright's Reamer: see *Reamer*.

Whittling Tray
The whittling tray was once a common item in many schools where elementary knife-work was taught. It was usually equipped with tools such as *compass, rule,* and *sloyd knife*.

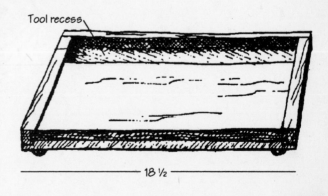

Tool recess

18 ½

Winding Sticks
Winding sticks, also called "winders," are two straight-edged sticks of equal width and length (20 in. to 30 in.) used for testing the *winding* or flatness of a board being planed. By laying them across opposite ends of a board and sighting across their tops any twist is easily apparent.

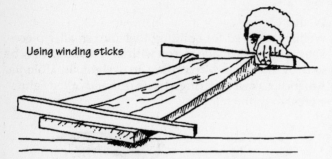

Using winding sticks

Wood Axe: see Axe.

Woodcarver's File: see *File*.

Woodcarver's Rasp: see *Rasp*.

Woodcarver's Vise: see *Vise*.

Woodcarving Chisel: see *Chisel, Wood, Woodcarving*.

Wood Chisel: see *Chisel, Wood*.

Wood Rasp: see *Rasp*.

Woodturning Chisel: see *Chisel, Wood, Woodturning*.

Woodworker's Vise: see *Vise*.

Wrecking Bar: see *Ripping Bar*.

Wrench (Br. Spanner)
A wrench used for tightening and loosening nuts and bolts is known in Britain as a spanner. There are hundreds of varieties, but the woodworker usually has at least one, of which the type illustrated is typical.

Adjustable crescent wrench

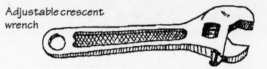

Yankee Screwdriver: see *Screwdriver*.

Yardstick: see *Rule*.

Zig-zag Rule: see *Rule*.

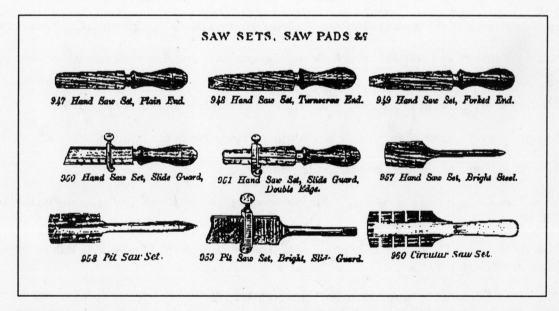

SAW SETS, SAW PADS &c

947 Hand Saw Set, Plain End.   948 Hand Saw Set, Turnscrew End.   949 Hand Saw Set, Forked End.

950 Hand Saw Set, Slide Guard,   951 Hand Saw Set, Slide Guard, Double Edge.   957 Hand Saw Set, Bright Steel.

958 Pit Saw Set.   959 Pit Saw Set, Bright, Slide Guard.   960 Circular Saw Set.

Saw Wrests, from *The Illustrated Price List of Woodworking Tools*, of Alexander Mathieson & Sons, Ltd., 1899

BIBLIOGRAPHY

Any bibliography on woodworking handtools with pretensions to completeness would be too long to include here. Instead, I have listed a few of my favorites, together with some of the recognized cornerstones of the subject. Many of these contain more exhaustive bibliographies, and the curious reader should have no difficulty pursuing the subject further, according to his or her particular interest.

Early American Industries Association. *Chronicle*. Williamsburg, VA. 1933.

Diderot, D., and D'Alembert, J. le R., eds. *Encyclopédie, ou Dictionnaire Raisonné des Sciences, des Arts et des Métiers*. Paris. 1772.

Goodman, W. L. *British Planemakers from 1700*. Arnold & Walker, Needham Market. 1978.

Goodman, W. L. *The History of Woodworking Tools*. G. Bell and Sons, Ltd., London. 1962.

Hibben, Thomas. *The Carpenter's Tool Chest*. J. B. Lippincott Company, Philadelphia. 1933.

Hummel, Charles F. *With Hammer in Hand*. University Press of Virginia, Charlottesville. 1968.

Industrial School Association, The. *How to Use Wood-Working Tools*. Ginn, Heath, & Co., Boston. 1884.

Kean, Herbert P., and Pollak, Emil S. *Collecting Antique Tools*. The Astragal Press, Morristown, NJ. 1990.

Mercer, Henry C. *Ancient Carpenters' Tools*. The Bucks County Historical Society. 1951.

Moxon, Joseph. *Mechanick Exercises or the Doctrine of Handy-Works*. London. 1703.

Nicholson, Peter. *Mechanical Exercise*. London. 1812.

Proudfoot, Christopher, and Walker, Philip. *Woodworking Tools*. Charles E. Tuttle Company, Rutland, VT. 1984.

Roberts, Kenneth D. *Some 19th Century English Woodworking Tools*. Ken Roberts Publishing Co., Fitzwilliam, NH. 1980.

Roubo, André-Jacob. *L'Art du Menuisier Ebéniste*. Paris. 1774.

Salaman, R. A. *Dictionary of Tools*. George Allen & Unwin, London. 1975.

Sloane, Eric. *A Museum of Early American Tools*. Wilfred Funk, Inc., New York. 1964.

Smith, Roger K. *Patented Transitional & Metallic Planes in America 1827–1927*. The North Village Publishing Co., Lancaster, MA. 1981.

Ulrey, Harry F. *Carpenters' and Builders' Library, Vol. I*. Howard W. Sams & Co., Indianapolis. 1970.

Welsh, Peter C. *Woodworking Tools 1600–1900*. U. S. Government Printing Office, Washington, D. C. 1969.

Wildung, Frank H. *Woodworking Tools at Shelburne Museum*. The Shelburne Museum, Shelburne, Vt. 1957.

COLLECTIONS

The Bucks County Historical Society. Doylestown, Pennsylvania.

Cooperstown Museum. Cooperstown, New York.

Deutsches Werkzeug Museum (German Tool Museum). Remscheid, Germany.

Maison de l'Outil (Tool Museum). Troyes, France.

The Old Mill Museum. East Meredith, New York.

Shelburne Museum. Shelburne, Vermont.

St. Albans City Museum. St. Albans, England

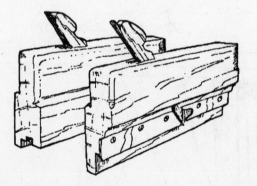